Spacetime and Geometry

Spacetime and Geometry

THE ALFRED SCHILD LECTURES

EDITED BY RICHARD A. MATZNER and L. C. SHEPLEY

UNIVERSITY OF TEXAS PRESS, AUSTIN

First Edition, 1982

Requests for permission to reproduce material from this work should be sent to Permissions, University of Texas Press, Box 7819, Austin, Texas 78712.

LIBRARY OF CONGRESS CATALOGING IN PUBLICATION DATA
Main entry under title:
Spacetime and geometry.
Includes index.
Contents: Why is the universe so symmetrical? / Dennis Sciama—Null congruences and Plebanski-Schild spaces / Ivor Robinson—Linearization stability / Dieter Brill—Nonlinear model field theories based on harmonic mappings / Charles W. Misner—[etc.]
1. General relativity (Physics)—Addresses, essays, lectures. 2. Gravitation—Addresses, essays, lectures. 3. Space and time—Addresses, essays, lectures. 4. Geometry—Addresses, essays, lectures. I. Schild, Alfred, 1921– . II. Shepley, Lawrence C., 1939– . III. Matzner, Richard A. (Richard Alfred), 1942– .
QC173.6.S67 530.1'1 81-11488
AACR2

ISBN: 978-0-292-74138-6

First paperback printing, 2012

To WINNIE SCHILD, Alfred's wife.

The greatness of a person reflects in his family, and Alfred's children and grandchildren are wonderful people. But Winnie is special.

Contents

Preface

These articles are meant to present current research topics at a level suitable for graduate students in physics. Each is based on a lecture from the Alfred Schild Memorial Lecture Series at the University of Texas at Austin. Alfred was a relativist of the first rank who especially made important contributions toward understanding geometrical aspects of the subject.

Alfred is remembered not just for his mathematical and physical works, but also for his personal attributes, including his unabashed admiration of great men of science and literature and his total support and enthusiasm for the research of his young colleagues. The essays in this volume keep with the tradition maintained by Alfred of bringing the brightest researchers to Texas and making them accessible to students. Thus, these essays have been written with pedagogy in mind. In some there are explicit exercises; in others the structure allows the reader to work through the development of the subject. The essays are presented here in the chronological order of the lectures on which they are based.

The authors and we ourselves are but a few of the people who loved and admired Alfred Schild. Alfred had enormous charm and wit and was knowledgeable of a wide range of topics. He was born in Istanbul (7 September 1921) and was educated in England and Canada, partly in internment camps during World War II (there are interesting stories here, of which we know rather little). He came to Austin in 1957 to the Department of Mathematics but became Ashbel Smith Professor of Physics in 1962, founding the Center for Relativity in that same year.

Alfred's best work included studies of the equivalence principle and geometry, work on quantum aspects of relativity, and studies of algebraically special and exact solutions. As important was his work as director of the Center for Relativity. In this role his idea of "directing" was simply to bring good people to Austin and give them full rein. The results were and continue to be tremendous; the research spans topics including black holes (the Kerr solution), kinetic theory, cosmology, radiation, algebraic structure, quantum theory, equations of motion, and many more.

Alfred's influence was also great in other aspects of physics, again with the philosophy of attracting good people and making sure they are free. He tirelessly strove to move the University of Texas toward real excellence and can truly be called the founder of theoretical physics at Austin.

Shortly after Alfred's sudden death (24 May 1977), a series of lectures in his honor was initiated. This volume is based on some of these lectures. The Alfred Schild Memorial Lecture Series continues to bring to Texas speakers, like the authors here, whom we admire not only as physicists but as people. In keeping with Alfred's philosophy, we give our lecturers full freedom. We particularly hope their personalities show in their contributions just as Alfred Schild's warm, generous nature appears in the work he did. Alfred is sorely missed.

Spacetime and Geometry

1. Why Is the Universe So Symmetrical?

DENNIS SCIAMA

In his introduction to this series of lectures, Ivor Robinson spoke for all of us about a fine scientist and an inspiring man. No one could have known Alfred for more than a few minutes without perceiving his unique blend of moral stature and lack of pretension.

His early death is a terrible tragedy. There is no way of coming to terms with it. All we can do is to try and carry on the same fight for moral standards, for lack of pretension, and, perhaps the greatest fight of all, to tear away from Nature the inscrutable veil that hides her inner workings. Alfred's weapons in this fight were inexorable logic, a finely tuned mathematical technique, and a deep feeling for the scientific beauty of Nature, or perhaps I should say for the beauty of our theories of Nature. That is partly why his two greatest scientific heroes were Albert Einstein and Paul Dirac, whom he regarded as supreme artists as well as supreme scientists.

But Alfred had another weapon, and with it he created at the University of Texas the Center for Relativity, which rapidly became the leading center in the whole of America. To Texas came relativists from all over the world, and from Texas came research in relativity that is admired all over the world.

To honor his memory this lecture will concern one aspect of his first major work. This work was carried out in collaboration with another of his scientific heroes, Léopold Infeld. It consisted of two papers entitled "A New Approach to Kinematical Cosmology," which were published in the *The Physical Review* in 1946. Their point of departure was E. A. Milne's attempt to do for the science of motion what Euclid had done for geometry, that is, to turn it into a string of theorems. Of course, strings of theorems need axioms, and Milne's main axiom was the cosmological principle that the large-scale structure of the universe is highly symmetrical. Infeld and Schild gave a penetrating analysis of the consequences of this axiom, an analysis that cannot be improved upon today.

What has changed since 1946 is the observational evidence. We know now that on a large scale the universe is far more symmetrical than even the most optimistic theorist would have expected. The main aims of this lecture are to discuss this evidence and to emphasize the problem to which it gives rise, namely, Why is the universe so symmetrical? We shall see that this problem is still unsolved.

Our first task must be to understand what we mean by "on a large scale." In more local problems we are familiar with this question. We know, for example, what it means to say that the Earth is a sphere, and we also know why it is a sphere. On closer inspection, of course, the Earth is not quite a sphere; for our later purposes the most relevant asymmetry arises from the Earth's rotation about its axis. The existence of this preferred direction destroys the isotropy of the Earth. Moreover, the dynamical consequences of the rotation—the flattening at the poles and the bulging at the equator—constitutes a further departure from isotropy. It will be relevant to our later discussion to ask, Relative to what is the Earth rotating? Newton's answer was absolute space. By contrast, Berkeley, Mach, and the early Einstein answered, relative to the fixed stars, or as we would now say, relative to the distant galaxies. We must refer to the distant galaxies because we now know that our own galaxy, the Milky Way, is itself rotating about a preferred axis, and that as a result it is (very) flattened at the poles and bulging at the equator. Hence the galaxy is rotating relative either to absolute space or to external matter.

When we turn to the whole universe we need to define a scale in terms of which we can discuss questions of symmetry. Such a scale is provided by the expansion of the universe that conforms very well (how well we must consider later) to the famous Hubble Law

$$v = \frac{r}{\tau}.$$

Here v is the radial velocity of a galaxy at a distance r. For small look-back times, τ is independent of r. We see immediately that τ gives us the scale we need since "small" means "much less than $c\tau$." The observed value of τ is close to 10^{10} years and is probably known to within a factor 2. Thus "large scale" means for times comparable to 10^{10} years or distances comparable to 10^{10} light years, that is, 3,000 megaparsecs (where relativistic effects would have to be included in the velocity-distance relation).

POSSIBLE LARGE-SCALE ASYMMETRIES OF THE UNIVERSE

Peculiar Velocities. We would not expect the motion of a particular galaxy to conform exactly to the Hubble Law. Of course, a galaxy, with its linear dimensions ~10 kiloparsecs, would represent a small-scale irregularity in the structure of the universe, analogous, say, to a mountain on the surface of the Earth. We are here more interested in the peculiar velocity of a cluster of galaxies (scale $\sim$1 Mpc) and still more in that of a supercluster (scale $\sim$50 Mpc), although even here we would be dealing with structures far smaller than the "radius of the universe" ($\sim$3,000 Mpc).

Inhomogeneities. We have just tacitly assumed that superclusters exist on a scale $\sim$50 Mpc. This does seem to be the consensus amongst galactic astronomers. Indeed, some superclusters stand out clearly in the general distribution of galaxies. However, for our purposes we are interested in the possibility of larger scale irregularities. Are there, for example, regions of significantly enhanced mean density (say, $\delta\rho/\rho \sim 0.1-1$) over length-scales $\sim$500–1,000 Mpc?

Vorticity. The question involved here is, Does the universe as a whole rotate relative to absolute space? This question is subtler than in the case of the Earth or the galaxy, because in those cases there is a unique axis of rotation passing through the center of the system. However, if the universe is homogeneous on a large scale it would have no center, and there would be a rotation axis through every point. The rotation of the universe would then manifest itself in the following way: By local dynamical experiments, using, say, a Foucault pendulum, one can determine a locally nonrotating frame of reference. If one is worried about local gravitational effects on the Foucault pendulum, such as would arise, say, at the surface of a rotating neutron star ("dragging of inertial frames"), one can determine instead a nonrotating frame on the scale of a galaxy or a cluster of galaxies, for example, the frame relative to which the Milky Way is rotating (and therefore flattened). If the whole system of galaxies is rotating relative to this frame, then we say that the universe is rotating relative to absolute space, and the angular velocity of the rotation is called the *vorticity* of the universe.

This is an oversimplified description of the procedure actually carried out in theoretical cosmology, but it gives the physical essence of what is involved. What one actually does in any given model of the universe is to expand the velocity field in the neighborhood of an observer in a Taylor series, just as one does in fluid dynamics. The vorticity corresponds to the flow that leads to an infinitesimal rotation of a set of fluid particles. In addition, there will be *expansion*, which just gives the Hubble effect in our case, and *shear*, which corresponds to a shape-changing transformation on the fluid particles and which we shall discuss next.

The possibility that the universe might have vorticity was pointed out in a famous paper by Gödel (1949) who gave an exact solution of Einstein's field equations for a nonexpanding rotating universe. This possibility disturbed Einstein, not so much for its anti-Machian character, since by then Einstein had lost his attachment to Mach's principle, but because the Gödel solution contained closed time-like lines, implying that one could enter into one's past.

An expanding universe can also have (anti-Machian) vorticity, and in that case the change of vorticity with time is determined by the law of conservation of angular momentum. In a matter-dominated universe of negligible pressure we would have for small ω,

$$\omega \propto \frac{1}{R^2},$$

where $R(t)$ is a measure of the distance between a pair of substratum particles. In a radiation-dominated universe ($p = \rho/3$), we would have

$$\omega \propto \frac{1}{R},$$

and in an ultra-stiff universe ($p = \rho$), we would have

$$\omega \propto R.$$

In this last case there is so much inertia in the matter field that eddies expanding with the universe actually spin up (Barrow, 1977). We would not expect an ultra-stiff equation of state to hold, except possibly in the very earliest stages of the expansion. Nevertheless, this possibility may play a key role in relation to the symmetry properties of the universe, as we shall see later.

Shear. We have seen that shear represents a shape-changing distortion arising from the velocity field in the neighborhood of a point. It would manifest itself cosmologically most directly in the Hubble constant τ being a function of direction, that is, the universe would expand at different rates in different directions. The velocity gradient involved has the dimensions of an inverse time (as does vorticity, of course). Qualitatively, the shear σ is given by

$$\sigma \sim \frac{\partial}{\partial \theta}\left(\frac{1}{\tau}\right),$$

where θ is a measure of direction, and the change of σ with time is given by

$$\sigma \propto \frac{1}{R^3}.$$

The shear and vorticity are important not only kinematically but also dynamically because they affect the rate of expansion of the universe. We can express this effect by attributing a positive energy to the shear and a negative energy to the vorticity (centrifugal effect). Einstein's field equations then imply that at early epochs the expansion rate $\dot{R}/R$ is given approximately by (Ellis, 1971; Olson, 1978)

$$3\left(\frac{\dot{R}}{R}\right)^2 \cong (8\pi G\rho + \sigma^2 - \omega^2).$$

Thus the shear-energy acts like matter to speed up the expansion, while the vorticity-energy slows it down. Note, however, how the different forms of energy change with time during the expansion. For pressure-free matter we would have

$$\rho \propto \frac{1}{R^3},$$

for radiation-dominated matter we would have

$$\rho \propto \frac{1}{R^4},$$

and for ultra-stiff matter we would have

$$\rho \propto \frac{1}{R^6}.$$

For vorticity-energy we would have

$$\omega^2 \propto \frac{1}{R^4} \quad (p = 0),$$

$$\propto \frac{1}{R^2} \quad \left(p = \frac{1}{3}\rho\right),$$

and

$$\propto R^2 \quad (p = \rho),$$

and for shear-energy we would have

$$\sigma^2 \propto \frac{1}{R^6}.$$

Thus if any large-scale shear at all is present it would dominate dynamically at sufficiently early times except in the ultra-stiff case.

OBSERVATIONAL LIMITS ON THE LARGE-SCALE ASYMMETRY OF THE UNIVERSE

Observations of Galaxies

Peculiar Velocities. A glance at the red shift–apparent magnitude relation for galaxies believed to be standard candles (Figure 1.1) shows that deviations from the Hubble Law do not exceed a few hundred kilometers per second ($v/c \sim 10^{-3}$). Individual velocities for galaxies in rich clusters can

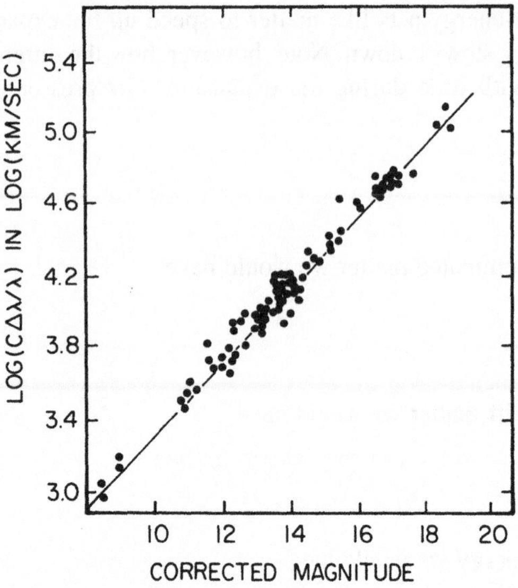

Figure 1.1. The magnitude-log red shift diagram based on the brightest galaxy in 89 clusters (after Sandage, 1972). The slope has the expected value of 5, required by all homogeneous models in the $z \rightarrow 0$ limit.

reach thousands of kilometers per second, but there is no evidence that a cluster as a whole can have a peculiar velocity as large as this.

In view of the interest attached to this question and the importance, in particular, of understanding the peculiar motion of our own galaxy, astronomers have made many studies of the local pattern of velocities, often with controversial results. Even the rotation velocity of the sun around the center of the Milky Way is uncertain to about 50 km sec^{-1}. (For a recent discussion, see Yahil, Tammann, and Sandage, 1977).

The motion of the Milky Way as a whole relative to the other galaxies in the local group can be derived from estimates of the proper motion of the sun relative to those galaxies by subtracting out the rotation of the sun around the center of our galaxy (de Vaucouleurs and Peters, 1968; de Vaucouleurs, Peters, and Corwin, 1977; Yahil, Tammann, and Sandage, 1977).

The main controversy arises in connection with the motion of the local group as a whole. This group appears to belong to a flattened supercluster of galaxies centered on the Virgo cluster (de Vaucouleurs, 1975). Analyses of our motion in the supercluster by various workers (de Vaucouleurs, 1958; Sciama, 1967; Stewart and Sciama, 1967; de Vaucouleurs and Peters, 1968) and of our motion relative to nearby galaxies outside the local group (Rubin et al., 1976; Schecter, 1977) have led to discordant results. We shall not, therefore, pursue this question here, but we recommend that the interested reader consult the literature cited, in which references to other discussions are given. We note only that the largest velocity derived for the local group is of order 600 km $\sec^{-1}$.

Inhomogeneities. We have seen that irregularities (superclusters) probably exist on a scale of 50 Mpc, but no convincing asymmetries have been established for galaxy distributions on a larger scale. In recent years there have been important developments in the study of correlations in the distribution of galaxies (Totsuji and Kihara, 1969; Fry and Peebles, 1978; Davis, Geller, and Huchra, 1978). (Earlier references are given in the last two papers.) This work is likely to give us valuable clues to the still poorly understood process of galaxy formation but is unlikely to yield useful information on the length-scale ~500–1,000 Mpc that is our concern here.

Vorticity. If the universe as a whole is rotating relative to our local inertial frame, we would expect to see a *transverse* Doppler shift in the spectra of distant galaxies, except for those lying in the direction of the rotation axis. Because the transverse effect is of second order in v/c, this method of detecting rotation is not very sensitive. The absence of any clear effect in the data of Figure 1.1 tells us only that $v \lesssim c$ at a Hubble radius (Kristian and Sachs, 1966). The rotation period P associated with such motion then clearly satisfies only the weak inequality

$$P = \frac{2\pi}{\omega} \gtrsim \tau .$$

Nevertheless, this is a more stringent limit on ω than one can deduce from the consideration that the flattening of the galaxy is compatible with observations of its differential rotation, which tells us that the outlying stars of the galaxy are as good as "fixed" stars for the purposes of this comparison. Since the orbital periods of these stars ~10^9 years, this tells us only that $P \gtrsim 10^9$ years, which is a ten times weaker limit than the Kristian-Sachs one. (Nevertheless, this limit is stronger than one could derive using the best available gyroscopes to determine the local nonrotating frame.)

Shear. The red shift–apparent magnitude relation of Figure 1.1 shows that there is no gross anisotropy in the value of the Hubble constant. As Kristian and Sachs (1966) pointed out, one obtains thereby only a weak limit on the shear of the same order as for the vorticity:

$$\frac{1}{\sigma} \gtrsim 3\tau .$$

Observations of the 3°K Cosmic Microwave Background

Peculiar Velocities. The motion of the Earth through the 3°K background can in principle be observed from the resulting Doppler effect. The spectrum in each direction remains that of a black body, but the temperature varies with the cosine of the angle between the directions of motion and of observation. Several attempts have been made to detect this effect, of which the most recent was by Smoot, Gorenstein, and Muller (1977). The observations were made at 33 GHz (0.9 cm) at an altitude of 20 km aboard a U-2 aircraft. At such a high frequency the emission from the galaxy would be unimportant. However, the data refer to only two-thirds of one hemisphere in the sky and were force-fitted to a first-order spherical harmonic. One cannot therefore be sure that the observed anisotropy is not due to a second-order harmonic. Fortunately, it is intended to extend the sky coverage to the other hemisphere.

Despite this uncertainty, the observed results must be taken seriously as a measurement of our velocity through the microwave background. The data are shown in Figure 1.2, and the fit to a first-order harmonic corresponds to a velocity of 390 ± 60 km $\sec^{-1}$ towards galactic longitude ~230° and latitude ~67°. When the sun's rotation in the galaxy of 300 ± 50 km $\sec^{-1}$ is subtracted out, one obtains for the velocity of the galaxy through the 3°K background

$$600 \text{ km sec}^{-1} \text{ towards } l \sim 260°,\ b \sim 39° .$$

This velocity is rather large but is in reasonable agreement with the earlier measurements of Corey and Wilkinson (1976), Henry (1971), and Conklin (1969). It does not agree with the results of Rubin et al. (1976) or de Vaucouleurs and Peters (1968) mentioned earlier (but see Muller, 1980).

The problem posed by this observation is indicated by the fact that it implies a substantial velocity transverse to the direction of the Virgo cluster, which lies close to the center of the local supercluster. At first sight this would be compatible with the idea that the supercluster is rotating and hence flattened. However, there are grave objections to this picture. First of all, the rotation period of the supercluster would be an order of magnitude greater than the age of the universe τ, so that there would not have been time for the "eddy" to flatten by relaxing. Second, in the absence of pressure gradients to stabilize the eddy, one cannot construct a sensible dynamical past history for it. Third, a cosmological distribution of such eddies would induce detectable distortions of the microwave background (Anile et al., 1976; Batakis and Cohen, 1975).

A more likely explanation for the peculiar velocity of 600 km $\sec^{-1}$ is that the local group is under the gravitational action of a nearby irregularity (supercluster) and that this velocity has been built up over the last 10^{10} years. This would require a gravitational acceleration of 2×10^{-10} cm $\sec^{-2}$,

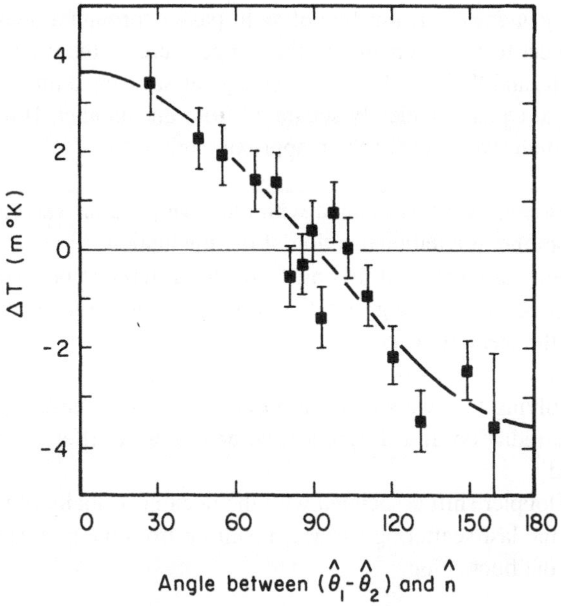

Figure 1.2. Measurement of the temperature anisotropy of the 2.7K background (after Muller, 1978). Apparatus consisted of two receivers with narrow (7°) beams observing regions of the sky 60° apart. The vector $\hat{\theta}_1 - \hat{\theta}_2$ is a unit vector pointing from the center of one horn to the center of the other. The unit vector $\hat{n}$ gives the direction toward the hottest region of the sky ($RA = 11.23 \pm 0.46$ hours, dec $= 19° \pm 7.5'$). As expected in a simple Doppler explanation of the anisotropy (only one hot region), the measured temperature difference between the two horns is proportional to $\hat{n} \cdot (\hat{\theta}_1 - \hat{\theta}_2)$ (i.e., a cosine wave). Because of the 60° geometry, the amplitude of the cosine variation is the same as the temperature excess on the sky, $3.61 \pm 0.54 \times 10^{-3}$K.

which would be exerted by a supercluster of $10^{16} M_{\odot}$ at a distance of 30 Mpc. It would be interesting to try and identify the supercluster that may be responsible.

An attempt has also been made to detect our velocity relative to the diffuse X-ray background, which is believed to originate in distant regions of the universe. This background has been observed to be isotropic, and the absence of a first harmonic angular distortion enables one to place an upper limit of about 800 km sec^{-1} on our velocity (Schwartz, 1970). This limit is clearly compatible with the microwave results, and it would be interesting if future X-ray observations could lead to a definite result that could be compared with the microwave data.

Inhomogeneities. If our line of sight passes through a density fluctuation, we expect to find a change in the temperature of the microwave background (Sachs and Wolfe, 1967). This change arises from a number of causes that in general cannot be clearly separated from one another. However, if the effects are small, we can split them approximately into

(α) a Doppler shift associated with our own peculiar velocity, resulting from the gravitational action of the fluctuation;

(β) a gravitational shift arising from the difference between the time-dependence of the gravitational potential of the fluctuation and that of the mean universe;

(γ) a time-delay associated with propagation through the fluctuation, implying that we would be observing the last-scattering surface of the radiation at a different time and so at a different temperature; and

(δ) a Doppler shift associated with the peculiar velocity of the electrons in the last-scattering surface, resulting from the gravitational action of the fluctuation.

The detailed calculation of all these effects is quite complicated, and in particular allowance must be made for the smearing produced by the finite depth of the last-scattering surface. Since no positive effect has been observed in the background, we shall not describe these calculations here but simply give an indication of the implied upper limits on possible large-scale density inhomogeneities. The interested reader should consult the papers of Sachs and Wolfe (1967), Rees and Sciama (1968), Dyer (1976), and Anile et al. (1976), in which references are given to other work.

Let us consider a particular concentration with density contrast 3 to 1 at a red shift $z \sim 1.5$, having a mass $\sim 3 \times 10^{19} M_{\odot}$. This concentration can be modeled in a zero-pressure Robertson-Walker universe by removing a comoving sphere of dust and replacing it with a smaller sphere of equal mass. This replacement would not affect the dynamics of the universe outside the sphere. In an Einstein–de Sitter universe the resulting hole would subtend an angle of about 20°, and its present diameter would be about 750 Mpc. According to Dyer, the center of such a hole would be about 0.2% colder than the surrounding universe away from the hole, while the radiation from the limb would be about 0.1% hotter. In a low-density universe (with deceleration parameter $q_0 \sim 0.05$) the central temperature would be about 0.1% hotter than in the surrounding universe.

Temperature fluctuations of this order of magnitude are not observed. If one subtracts out the cosine variation of Figure 1.2, the background is isotropic to 1 part in 3,000 (Smoot, Gorenstein, and Muller, 1977). Of course, 10% density fluctuations on the same length-scale could not be ruled out by these observations. This gives us the order of magnitude of the permitted large-scale irregularities in the distribution of matter in the universe (unless more refined observations succeed in reducing even further the upper limit on localized anisotropies in the microwave background).

Another type of inhomogeneity would arise if an observer's four-velocity is not orthogonal to the surfaces of homogeneity. This effect has been considered by King and Ellis (1973), Ellis and King (1974), Batakis and Cohen (1975), and Barrow and Tipler (1979).

Vorticity. If the universe had vorticity, the last-scattering surface of the microwave background would be rotating around us, giving rise to a transverse Doppler effect whose magnitude would depend on the angle between the direction of observation and the rotation axis. This question has been analyzed by Hawking (1969), Collins and Hawking (1973*a*), and Batakis and Cohen (1975). The results depend somewhat on the type of model universe one considers. If one makes the simplifying assumption that the universe is exactly homogeneous but anisotropic, one obtains generalizations of the Robertson-Walker models that can be classified in terms of the geometry of the homogeneous three-dimensional spaces of constant cosmic epoch. This classification was carried out by Bianchi in the last century, in terms of the symmetry properties of the three-dimensional spaces. There are nine Bianchi types, of which five admit Robertson-Walker models as special cases. In the usual notation (Ellis and MacCallum, 1969) these are

$$\begin{array}{lll} \mathrm{I},\ \mathrm{VII}_0 & \text{containing} & k = 0 \\ \mathrm{V},\ \mathrm{VII}_h & & k = -1 \\ \mathrm{IX} & & k = +1 . \end{array}$$

Homogeneous models of type I have zero vorticity. For the others, Collins and Hawking obtain the following limits (the limit in type VII_0 depends also on the comoving length scale over which the vorticity changes and is given in Collins and Hawking, 1973*a*):

$$\begin{array}{lll} \mathrm{V},\ \mathrm{VII}_h & \dfrac{2\pi}{\omega} > 1.5 \times 10^3 \tau & \\ \mathrm{IX} & 4 \times 10^8 \tau \ , & z_s \sim 7 \\ & 6 \times 10^{11} \tau \ , & z_s \sim 1{,}000 . \end{array}$$

Here z_s is the red shift of the last-scattering surface; z_s would be of order 7 in a high-density universe most of whose mass is in the form of ionized intergalactic gas and of order 1,000 otherwise. At the moment the latter seems the more plausible.

The increase in the stringency of these limits is very marked. However, one must remember that during the matter-dominated phase of the expansion the ratio of rotation period to expansion time scale increases with the age of the universe. To some extent, then, the large value of this ratio at the present epoch simply reflects the advanced age of the universe today. If we want to say that the universe is rotating slowly, or not at all, we must consider what limits can be placed on the total number of rotation periods that may have occurred since the big bang. This question was also considered by Collins and Hawking (1973*a*). Their strongest limit arises in type IX (closed universe), for which they found that the universe could have rotated through only 2×10^{-4} seconds of arc since the big bang.

Shear. It is clear that if the universe expands at a different rate in different directions the microwave background would have a different temperature in different directions. This question was also analyzed by Collins and Hawking (1973*a*), who arrived at the following limits:

I	$\dfrac{1}{\sigma} >$	$10^4\tau$,	$z_s \sim 7$
		$1.4 \times 10^7\tau$,	$z_s \sim 10^3$
V		$7 \times 10^3\tau$	
VII_0		$7 \times 10^2\tau$	
VII_h		$7 \times 10^3\tau$	
IX		$10^3\tau$.	

Again, there is a considerable improvement on the limits that can be derived from observations of galaxies.

Observations of the Cosmic Helium Abundance

If we accept that most of the helium we observe was formed by thermonuclear processes 100 seconds after the hot big bang (Reeves, 1974), we can determine the expansion time-scale prevailing during the nuclear reactions (assuming that the lepton number was small). We know that the Robertson-Walker time-scale leads to agreement with observation, so that the uncertainty in the observations provides us with an upper limit on the large-scale shear and vorticity of the universe, since as we saw earlier the existence of such anisotropies leads to a change in the rate of expansion. We also saw that the shear energy increases into the past much faster than the vorticity energy, so that in practice the limits one can place on the vorticity are much weaker, weaker indeed than those derived from the isotropy of the microwave background.

The reverse is true for the shear, as was first pointed out by Barrow

(1976). A modified calculation has more recently been presented by Olson (1978). The various Bianchi types that admit Robertson-Walker models can be taken together, except one must exclude the theoretical possibility of growing shear modes, which would be allowed in types VII_h and IX. If we take from observation that the helium to hydrogen ratio by mass lies between 0.25 and 0.33, then we obtain a strong upper limit for the present value of the large-scale shear, namely

$$\frac{1}{\sigma} \gtrsim 5 \times 10^{11} \tau .$$

This limit is four orders of magnitude more stringent than the best limit derivable from the isotropy of the microwave background.

Observations of the Entropy in the Microwave Background

As originally emphasized by Misner (1968) an important aspect of any large-scale asymmetry in the universe is the possibility that it has been reduced by dissipation. Associated with this dissipation would be the production of heat, and this heat in turn would contribute to the general radiation background. At centimeter and millimeter wavelengths this background is observed to have an approximately thermal spectrum. Such a spectrum would be produced as a result of interaction with matter that is at a definite temperature. But to guarantee thermalization on the available time-scale, this interaction must have occurred when the universe was denser than it is today by a factor of at least about 10^{15}. Such early dissipation would then contribute to the entropy of radiation in the universe. A convenient measure of this entropy is the entropy per baryon or, the equivalent, the number of photons per baryon, inasmuch as this quantity is time-independent in an expanding universe in the absence of further dissipation. The value of n_{ph}/n_b today is about 10^8 in an Einstein–de Sitter universe.

This ratio was originally viewed as a large number, to be attributed either to initial conditions at the hot big bang or to the occurrence of substantial dissipation at early times, mostly before 100 seconds (if thermonuclear reactions at that time are to be responsible for the observed abundance of helium). However, it has recently been emphasized by Barrow and Matzner (1977) that the ratio of 10^8 may represent a very *small* number. The reason for this is that, as we have seen, the shear energy increases faster into the past than the radiation energy. Hence, the earlier the shear is dissipated the more entropy per baryon is produced.

We are therefore impaled on what might be called the Barrow-Matzner fork. If the dissipation rate is low, the early shear must be small so that the observed abundance of helium is obtained. If the dissipation rate is large, the early shear must be small to ensure that not more than 10^8 photons per baryon are produced. If, for example, the dissipation occurred at or before a time ~1 second, then at present

$$\frac{1}{\sigma} \gtrsim 10^{13}\tau,$$

which is a 20 times stronger limit than the one derived directly from the helium argument.

This argument also limits the initial shear of the universe. If, say, there was strong dissipation due to quantum processes at 10^{-43} seconds (Zel'dovich, 1972), the initial shear would have had to be so small that even without dissipation we would call it essentially zero (the undissipated shear-energy density today would be less than 10^{-39} of the matter energy density (Barrow and Matzner, 1977).

WHY IS THE UNIVERSE SO SYMMETRICAL?

The upshot of our discussion so far is that, in all probability, the universe has always been close to the exactly homogeneous and isotropic state characterized by the Robertson-Walker models. We must now face the question, Why is the universe so symmetrical? A further question is, Why do portions of the last-scattering surface of the microwave background lying in widely different directions have so precisely the same temperature ($\Delta T/T < 3 \times 10^{-4}$) when in a Robertson-Walker model such portions would not have had time to communicate with one another?

Four possible answers have been given to the first question:

1. initial conditions
2. anthropic principle
3. ultra-stiff early universe
4. Mach's principle

We consider these answers in turn.

Initial Conditions

Penrose has suggested (1978, 1979) that the initial symmetry of the universe is related to the second law of thermodynamics. He points out that a gravitational field has entropy when it is irregular, due to its negative specific heat (Lynden-Bell and Lynden-Bell, 1977).

He pictures the universe as developing from an initially regular state to a state in which considerable entropy develops in the gravitational field. He characterizes the initially regular state by the vanishing of the Weyl tensor, which corresponds essentially to the universe being initially exactly like the Robertson-Walker model. Since the final state of the universe is not gravitationally regular, one would be dealing here with a time asymmetrical initial condition. Penrose relates this condition to the existence of a local microscopic law that is noninvariant under both time reversal T and charge conjugation, space inversion, and time reversal (CPT). This is an interesting suggestion, but the relation between the noninvariant law and the vanishing of the Weyl tensor needs further elucidation.

This idea may also be related to the problem of the presumed dominance of matter over antimatter in the universe. The development of irregularities may lead to the formation of black holes, a process that would be asymmetrical in time. These black holes may radiate more baryons than antibaryons by the Hawking process (Hawking 1974, 1975) if there exists an appropriate microscopic interaction that is not invariant under C, P, and T (Toussaint et al., 1979).

Anthropic Principle

The anthropic principle has been invoked in this context by Collins and Hawking (1973*b*). They argue that galaxies can form only in an isotropic universe, and therefore if human beings exist they must find the universe to be isotropic. To quote Collins and Hawking: "The answer to the question why is the Universe isotropic appears to be, because we are here."

This prompts the further question, Why are we here? If our existence requires a carefully contrived universe, why did the universe have this careful contrivance? An extreme form of the anthropic principle answers this question by invoking the existence of all conceivable universes. Intelligent life would then develop in some subset of these universes that permits such a development. Such intelligent life would in turn find present in its universe all those features that are needed for its development. The existence of these features would then not require further explanation (Carter, 1974).

I have considerable sympathy for this extreme form of the anthropic principle. However, its application to the isotropy of the universe must be regarded as uncertain until it can be shown in detail that the formation of galaxies does depend critically on this isotropy. We understand so little about galaxy formation (Jones, 1976) that this must be regarded as an open question.

Ultra-stiff Early Universe

It has recently been pointed out by Barrow (1977, 1978) that if the equation of state for baryons in the early universe is of the ultra-stiff kind, characterized by

$$p = \rho ,$$

one might be able to understand the low asymmetry of the early universe. For such an extreme equation of state, the time-dependence of any asymmetry is quite different from that holding for more conventional equations of state. We saw an example of this in our first discussion of vorticity: in the ultra-stiff regime, eddies (and the whole universe) *spin up* as the universe expands. As we shall see, similar differences occur for the other kinds of asymmetry.

The idea that at supernuclear densities the equation of state for baryons becomes ultra-stiff was first proposed by Zel'dovich (1962). The model that he had in mind was that of a population of stationary baryons whose interaction is mediated by a spin 1 field. This model was improved by Walecka (1974, 1975), who used relativistic many-body quantum field theory. Further considerations in favor of the ultra-stiff equation of state have been advanced by Canuto (1974, 1975). For such an equation of state it is relatively easy for dissipation to produce a significant amount of entropy (Liang, 1977).

We now consider the effect of this equation of state on the various kinds of asymmetry in the universe.

Peculiar Velocities. In the early universe we have in general

$$\rho \propto \frac{1}{t^2},$$

and in the ultra-stiff case ($p = \rho$)

$$v \propto t^{2/3}.$$

By contrast, in a radiation-dominated universe ($p = \rho/3$)

$$v \propto t^0,$$

and in a pressure-free matter-dominated universe ($p = 0$),

$$v \propto t^{-2/3}.$$

Thus any peculiar velocity would diverge initially unless the equation of state is as stiff as that for the conventional radiation-dominated case. In fact for

$$p = (\gamma - 1)\rho,$$

we would have

$$v \propto t^{-\frac{2}{\gamma}\left(\frac{4}{3} - \gamma\right)}.$$

Inhomogeneities. For simplicity we consider here inhomogeneities so large that their length-scale exceeds the distance to the particle horizon ct. Then the time-dependence of the density contrast $\delta\rho/\rho$ is given by (Liang, 1975*a*):

$$\frac{\delta\rho}{\rho} = A(\mathbf{x})t^{(\gamma-2)/2} + B(\mathbf{x})t^{2(3\gamma-2)/(3\gamma)}.$$

For $\gamma < 2$, the A perturbation decreases with time and the B perturbation increases with time. The increasing perturbation has been discussed in connection with the formation of galaxies, although its algebraic rather than exponential dependence on time leads to difficulties. The decreasing perturbation diverges at the origin and is normally excluded ad hoc. However, as Barrow (1978) points out, for the ultra-stiff case $\gamma = 2$, this perturbation no longer diverges. In addition, no shocks can occur because the velocity of sound equals the velocity of light; therefore, no entropy is produced by nonlinear hydrodynamical processes (Liang, 1975*b*).

Vorticity. Conservation of angular momentum implies that the vorticity is given by

$$\omega \propto t^{2(3\gamma - 5)/(3\gamma)} .$$

Thus for $\gamma > 5/3$ the vorticity tends to zero as t tends to zero. In the next section we shall discuss a deeper reason why the large-scale vorticity should be zero initially (and at all later times).

Shear. We have already seen that the large-scale shear σ is given by

$$\sigma \propto \frac{1}{R^3} .$$

Now for the density ρ we have

$$\rho \propto \frac{1}{R^{3\gamma}} ,$$

so that the *relative* shear energy density σ^2/ρ is given by

$$\frac{\sigma^2}{\rho} \propto R^{3\gamma - 6} .$$

Thus precisely for $\gamma = 2$, the shear energy density scales with the matter energy density, so that a "general" amount of shear that is dissipated early on need not lead to more entropy than is observed. In fact the shear energy associated with inhomogeneities of amplitude $\delta\rho/\rho \sim 10^{-4}$, which becomes dissipated after entering the horizon at a time $\sim 10^{-23}$ seconds, would account for an entropy ratio of 10^8 photons per baryon at the ultra-stiff era (Zel'dovich, 1972). Moreover, similar inhomogeneities on a larger scale could account for the later formation of galaxies (Zel'dovich, 1972). However, the entropy ratio produced depends sensitively on the epoch of dissipation.

Mach's Principle

We have already seen that, according to Mach's principle, local inertial frames are determined by the large-scale distribution of matter in the universe, so that there should be no net vorticity. As pointed out by Barrow (1977), if $\gamma = 2$ the universe would be unstable to the development of localized vorticity even if the total vorticity remains zero. Such localized vorticity might be associated with the subsequent formation of galaxies.

In order to investigate the further consequences of Mach's principle, one needs a precise mathematical formulation of the principle. This has been given by Raine (1975) following earlier work by Al'tshuler (1967); Lynden-Bell (1967); and Sciama, Waylen, and Gilman (1969). In this formulation one selects boundary conditions for solutions of Einstein's field equations that guarantee that the metric of spacetime is determined by its contents, with no contribution from "absolute space." The difficulty of the problem is that Einstein's equations are nonlinear, so that it is not clear what "determined by" means. Raine offers a solution to this problem, but it is inevitably rather complicated so we shall not describe it here but refer the reader to the original paper. The result that is relevant for our discussion is that a spatially homogeneous universe (that is, one of the Bianchi types) whose contents possess a perfect fluid equation of state and that is Machian in the sense of Raine must have zero vorticity and zero shear, so that in fact it conforms to the Robertson-Walker model.

However, what is left unexplained by this procedure is the large-scale homogeneity of the universe, which does not seem to be demanded by Raine's form of Mach's principle. In fact, it is important that the principle should not demand exact homogeneity or it would be in trouble with observation, and Raine does show that his theory passes this test. One would expect that a Machian universe with small-scale inhomogeneity would possess only the anisotropy needed to permit the inhomogeneity to exist, but a precise formulation of this restriction remains to be discovered.

Finally, we consider the question of why regions on the last-scattering surface of the microwave background, which in an isotropic universe would never have communicated with one another, manage to have the same temperature to such precision ($\Delta T/T < 3 \times 10^{-4}$). If dissipation of some kind produced the heat, why should the irregularities involved have such similar amplitudes in unrelated parts of the universe? Misner (1968) proposed an ingenious solution to this problem by invoking a "mixmaster" universe, in which early deviations from the Robertson-Walker model led to the elimination of particle horizons, permitting transport processes to even out the temperature. Subsequent work, however, showed that model universes with this property were still very special, so that it is just as arbitrary to invoke those models as it would be to invoke a uniform temperature in the first place. In addition, we have now seen that, in all probability, the universe has never deviated much from the Robertson-Walker state.

One far-reaching possibility is to consider that there was a collapse phase of the universe preceding the big bang, that it was during the late stages of this collapse that the entropy in the microwave background was generated, and that transport processes evened out the entropy during this phase. Such a possibility would require the singularity theorems for the big bang (Hawking and Ellis, 1973) to be evaded. One way in which this might be done, which is under active investigation at the moment, is to invoke quantum gravitational effects at the Planck time $\sim 10^{-43}$ seconds. One would expect the quantum aspects of the stress-energy tensor and of the gravitational field to be important at that time, and one knows from studies of similar processes near black holes (Candelas and Sciama, 1977) that these quantum effects can give rise to effectively negative energies, thereby reducing the strength of the gravitational field. In this way one might be able to get round the singularity theorems; technically one would be denying the validity of the energy condition that is so essential to this proof. It is not yet known whether this scheme will work. If it does not, the isotropy of the microwave background will remain to underline how little we understand about the basic processes that determine the large-scale symmetry of the universe.

ACKNOWLEDGMENTS

I am grateful to E. P. T. Liang for his helpful comments and to J. D. Barrow for his detailed criticisms of an early draft of this article and for guiding me through the intricacies of modern thinking about the symmetry of the universe.

NOTE

This lecture was delivered 28 October 1977. Since then, several observations of the properties of the 3°K cosmic background radiation have been reported. Features seen in the background are small effects and consistent with Professor Sciama's picture of the regularity of the universe. Besides the dipole anisotropy (galactic motion ~600 km/sec through the microwave background) (Smoot, Gorenstein, and Muller, 1977), there have been reports of detection of quadrupole anisotropy in the infrared ($\sim 10^{-3}$°K) by Fabbri et al. (1980) and in the microwave ($\sim 0.5 \times 10^{-3}$°K) by Boughn, Cheng, and Wilkinson (1981). The quadrupole variation may be explained in terms of the clustering of galaxies (Peebles, 1981). A search for polarization in the microwave has produced strong upper limits on the linear polarization ($\lesssim 6 \times 10^{-3}$°K at 0.91-cm wavelength) (Lubin and Smoot, 1981). This result implies strict limits in standard models on the amount of anisotropic expansion. Finally, Warwick, Pye, and Fabian (1980) report ~1% intensity variation in the X-radiation background. This may be consistent with large-scale (but small amplitude) inhomogeneities or with a very low-density, spatially homogeneous cosmology (Matzner, 1980)—The Editors, January 1981.

REFERENCES

Al'tshuler, B. L., "Integral Form of the Einstein Equations and a Covariant Formulation of Mach's Principle." *Sov. Phys.-JETP* **24**, 766–771 (1967).

Anile, A. M., Danese, L., De Zotti, G., and Motta, S., "Cosmological Turbulence Reexamined," *Astrophys. J. Lett. Ed.* **205**, L59–L63 (1976).

Barrow, J. D., "Light Elements and the Isotropy of the Universe," *Mon. Not. R. Astron. Soc.* **175**, 359–370 (1976).

———, "On the Origin of Cosmic Turbulence," *Mon. Not. R. Astron. Soc.* **179**, 47P–49P (1977).

———, "Quiescent Cosmology," *Nature* **272**, 211–215 (1978).

Barrow, J. D., and Matzner, R. A., "The Homogeneity and Isotropy of the Universe," *Mon. Not. R. Astron. Soc.* **181**, 719–727 (1977).

Barrow, J. D., and Tipler, F., "Analysis of the Generic Singularity Studies by Belinskii, Khalatnikov, and Lifshitz," *Phys. Rep.* **56**, 371–402 (1979).

Batakis, N., and Cohen, J. M., "Cosmological Model with Expansion, Shear, and Vorticity," *Phys. Rev.* **D12**, 1544–1550 (1975).

Boughn, S., Cheng, E. S., and Wilkinson, D. T., "Dipole and Quadrupole Anisotropy of the 2.7K Radiation," *Astrophys. J. Lett.* **243**, L113–L117 (1981).

Candelas, P., and Sciama, D. W., "Irreversible Thermodynamics of Black Holes," *Phys. Rev. Lett.* **38**, 1372–1375 (1977).

Canuto, V., "Equation of State at Ultrahigh Densities: Part 1," *Ann. Rev. Astron. & Astrophys.* **12**, 167–214 (1974).

———, "Equation of State at Ultrahigh Densities: Part 2," *Ann. Rev. Astron. & Astrophys.* **13**, 335–380 (1975).

Carter, B., "Large Number Coincidences and the Anthropic Principle in Cosmology," in *Confrontation of Cosmological Theories with Observational Data* (edited by M. S. Longair, Reidel, 1974), pp. 291–298.

Collins, C. B., and Hawking, S. W., "The Rotation and Distortion of the Universe," *Mon. Not. R. Astron. Soc.* **162**, 307–320 (1973*a*).

———, "Why Is the Universe Isotropic?" *Astrophys. J.* **180**, 317–334 (1973*b*).

Conklin, E. K., "Velocity of the Earth with Respect to the Cosmic Background Radiation," *Nature* **222**, 971–972 (1969).

Corey, B. E., and Wilkinson, D. T., "A Measurement of the Cosmic Microwave Background Anisotropy at 19 GHz," *Bull. Am. Astron. Soc.* **8**, 351 (1976).

Davis, M., Geller, M. J., and Huchra, J., "The Local Mean Mass Density of the Universe: New Methods for Studying Galaxy Clustering," *Astrophys. J.* **221**, 1–18 (1978).

de Vaucouleurs, G., "Further Evidence for a Local Super-cluster of Galaxies: Rotation and Expansion," *Astron. J.* **63**, 253–266 (1958).

———, "Supergalactic Studies: I. Supergalactic Distribution of the Nearest Galaxies," *Astrophys. J.* **202**, 319–326 (1975).

de Vaucouleurs, G., and Peters, W. L., "Motion of the Sun with Respect to the Galaxies and the Kinematics of the Local Supercluster," *Nature* **220**, 868–874 (1968).

de Vaucouleurs, G., Peters, W. L., and Corwin, H. G., "Possible New Members of the Local Group of Galaxies from Solar Motion Solutions," *Astrophys. J.* **211**, 319–323 (1977).

Dyer, C. C., "The Gravitational Perturbation of the Cosmic Background Radiation by Density Concentrations," *Mon. Not. R. Astron. Soc.* **175**, 429–447 (1976).

Ellis, G. F. R., "Relativistic Cosmology," in *General Relativity and Cosmology* (edited by R. K. Sachs, Academic Press, New York, 1971), pp. 104–182.

Ellis, G. F. R., and King, A. R., "Was the Big Bang a Whimper," *Commun. Math. Phys.* **38**, 119–156 (1974).

Ellis, G. F. R., and MacCallum, M. A. H., "A Class of Homogeneous Cosmological Models," *Commun. Math. Phys.* **12**, 108–141 (1969).

Fabbri, R., Guidi, I., Melchiori, F., and Natale, V., "Measurement of the Cosmic-Background Large-scale Anisotropy in the Millimetric Range," *Phys. Rev. Lett.* **44**, 1563–1566 (1980).

Fry, J. N., and Peebles, P. J. E., "Statistical Analysis of Catalogs of Extragalactic Objects: IX. The Four-point Galaxy Correlation Function," *Astrophys. J.* **221**, 19–33 (1978).

Gödel, K., "An Example of a New Type of Cosmological Solution of Einstein's Field Equations of Gravitation," *Rev. Mod. Phys.* **21**, 447–450 (1949).

Hawking, S. W., "On the Rotation of the Universe," *Mon. Not. Roy. Astron. Soc.* **142**, 129–141 (1969).

———, "Black Hole Explosions?" *Nature* **248**, 30–31 (1974).

———, "Particle Creation by Black Holes," *Commun. Math. Phys.* **43**, 199–220 (1975), and in *Quantum Gravity* (edited by C. J. Isham, R. Penrose, and D. W. Sciama, Oxford, 1975), pp. 219–267.

Hawking, S. W., and Ellis, G. F. R., *The Large Scale Structure of Space-time* (Cambridge Univ. Press, 1973).

Henry, P., "Isotropy of the 3K Background," *Nature* **231**, 516–518 (1971).

Jones, B. J. T., "The Origin of Galaxies: A Review of Recent Theoretical Developments and Their Confrontation with Observation," *Rev. Mod. Phys.* **48**, 107–149 (1976).

King, A. R., and Ellis, G. F. R., "Tilted Homogeneous Cosmological Models," *Commun. Math. Phys.* **31**, 209–242 (1973).

Kristian, J., and Sachs, R. K., "Observations in Cosmology," *Astrophys. J.* **143**, 379–399 (1966).

Liang, E. P. T., "Anisotropy and Large Scale Density Inhomogeneity in Nonrotating Cosmologies," *Phys. Lett. A* **51**, 141–143 (1975*a*).

———, "A Note on the Zel'dovich $p = \rho$ Cold Big Bang," *Mon. Not. R. Astron. Soc.* **171**, 551–553 (1975*b*).

———, "Entropy Generation in the Very Early Universe," *Phys. Rev. D* **16**, 3369–3375 (1977).

Lubin, P. M., and Smoot, G. F., "Polarization of the Cosmic Background Radiation," Lawrence Berkeley Laboratory Reprint (1981).

Lynden-Bell, D., "On the Origins of Space-time and Inertia," *Mon. Not. R. Astron. Soc.* **135**, 413–428 (1967).

Lynden-Bell, D., and Lynden-Bell, R. M., "On the Negative Specific Heat Paradox," *Mon. Not. R. Astron. Soc.* **181**, 405–419 (1977).
Matzner, R. A., "On Observations of the Cosmic Radiation Background," *Astrophy. J.* **241**, 851–857 (1980).
Misner, C. W., "The Isotropy of the Universe," *Astrophys. J.* **151**, 431–457 (1968).
Muller, R. A., "The Cosmic Background Radiation and the New Aether Drift," *Sci. Am.* **238**, No. 5, 64–74 (1978).
———, "Cosmic Microwave Background—Present Status and Future Prospects," in *Ninth Texas Symposium on Relativistic Astrophysics* (a symposium held in Munich, December 14–19, 1978, edited by J. Ehlers, J. Perry, and M. Waller) *Annals of the New York Academy of Sciences* **336**, 116–124 (1980).
Olson, D. W., "Helium Production and Limits on the Anisotropy of the Universe," *Astrophys. J.* **219**, 777–780 (1978).
Peebles, P. J. E., "Large-scale Fluctuations in the Microwave Background and the Small-scale Clustering of Galaxies," *Astrophys. J. Lett.* **243**, L119–L122 (1981).
Penrose, R., "Singularities of Spacetime," in *Theoretical Principles in Astrophysics and Relativity* (edited by N. R. Lebovitz, W. H. Reid, and P. O. Vandervoort, Univ. of Chicago, 1978), pp. 217–243.
———, "Singularities and Time-Asymmetry," in *General Relativity: An Einstein Centenary Survey* (edited by S. W. Hawking and W. Israel, Cambridge Univ. Press, 1979), pp. 581–638.
Raine, D. J., "Mach's Principle in General Relativity," *Mon. Not. R. Astron. Soc.* **171**, 507–528 (1975).
Rees, M. J., and Sciama, D. W., "Large-scale Density Inhomogeneities in the Universe," *Nature* **217**, 511–516 (1968).
Reeves, H., "On the Origin of the Light Elements," *Ann. Rev. Astron & Astrophys.* **12**, 437–467 (1974).
Rubin, V. C., Thonnard, N., Ford, W. K., and Roberts, M. S., "Motion of the Galaxy and the Local Group Determined from the Velocity Anisotropy of Distant Sc I Galaxies: II. The Analysis for the Motion," *Astron J.* **81**, 719–737 (1976).
Sachs, R. K., and Wolfe, A. M., "Perturbations of a Cosmological Model and Angular Variations of the Microwave Background," *Astrophys. J.* **147**, 73–90 (1967).
Sandage, A., "The Redshift-Distance Relation. II. The Hubble Diagram and Its Scatter for First-ranked Cluster Galaxies: A Formal Value for q_0," *Astrophys. J.* **178**, 1–24 (1972).
Schecter, P. L., "On the Solar Motion with Respect to External Galaxies," *Astron J.* **82**, 569–576 (1977).
Schwartz, D. A., "The Isotropy of the Diffuse Cosmic X-rays Determined by oso-III," *Astrophys. J.* **162**, 439–444 (1970).
Sciama, D. W., "Peculiar Velocity of the Sun and the Cosmic Microwave Background," *Phys. Rev. Lett.* **18**, 1065–1067 (1967).
Sciama, D. W., Waylen, P. C., and Gilman, R. C., "Generally Covariant Integral Formulation of Einstein's Field Equations," *Phys. Rev.* **187**, 1762–1766 (1969).
Smoot, G. F., Gorenstein, M. V., and Muller, R. A., "Detection of Anisotropy in the Cosmic Blackbody Spectrum," *Phys. Rev. Lett.* **39**, 898–901 (1977).

Smoot, G. F., and Lubin, P. M., *Astrophys. J. Lett. Ed.* **234**, L83 (1979).

Stewart, J. M., and Sciama, D. W., "Peculiar Velocity of the Sun and Its Relation to the Cosmic Microwave Background," *Nature* **216**, 748–753 (1967).

Totsuji, H., and Kihara, T., "The Correlation Function for the Distribution of Galaxies," *Publ. Astron. Soc. Jpn.* **21**, 221–229 (1969).

Toussaint, D., Treiman, S. B., Wilczek, F., and Zee, A., "Matter-antimatter Accounting, Thermodynamics, and Black Hole Radiation," *Phys. Rev.* **D19**, 1036–1045 (1979).

Walecka, J. D., "A Theory of Highly Condensed Matter," *Ann. Phys. (USA)* **83**, 491–529 (1974).

———, "Equation of State for Neutron Matter at Finite T in a Relativistic Mean-field Theory," *Phys. Lett. B* **59**, 109–112 (1975).

Warwick, R. S., Pye, J. P., and Fabian, A. L., "The Isotropy of the X-ray Background in the Energy Range 2-18 keV," *Mon. Not. Roy. Astron. Soc.* **190**, 243–260 (1980).

Yahil, A., Tammann, G. A., and Sandage, A., "The Local Group: The Solar Motion Relative to Its Centroid," *Astrophys. J.* **217**, 903–915 (1977).

Zel'dovich, Y. B., "The Equation of State at Ultrahigh Densities and Its Relativistic Limitations," *Sov. Phys.-JETP* **14**, 1143–1147 (1962).

———, "A Hypothesis, Unifying the Structure and the Entropy of the Universe," *Mon. Not. R. Astron. Soc.* **160**, 1P–3P (1972).

2. Null Congruences and Plebanski-Schild Spaces

IVOR ROBINSON

1. INTRODUCTION

More than half a century ago, Cartan (1922) observed that there were four null directions associated with the Weyl tensor. These directions, neglected for many years, have played an important part in the modern theory of exact solutions. My own involvement began in 1956 with an observation on null gravitational fields. There the four directions coincide; and a vector l_a in the common direction satisfies

$$l_a C^{abcd} = 0 . \tag{1.1}$$

In empty space, however,

$$C^{abcd}{}_{;d} = 0 , \tag{1.2}$$

from the Bianchi identities. Hence,

$$l_{a;d} C^{abcd} = 0 . \tag{1.3}$$

The first two equations remain true if the Weyl tensor is replaced by its dual (defined as in equation 2.7 below). Consequently,

$$l_{a;d} {}^*C^{abcd} = 0 . \tag{1.4}$$

Equations (1.3) and (1.4) are invariant under transformations of the form

$$C_{abcd} \to \lambda C_{abcd} + \mu {}^*C_{abcd} . \tag{1.5}$$

But this is precisely the extent to which the principal null directions determine a Weyl tensor. The appearance of this tensor in (1.3) and (1.4) is purely incidental; therefore, the essence of the equations is a restriction on the vector field. This restriction turns out to be

$$l_{a;b} l^b = 0 , \qquad l_{(a;b)} l^{a;b} = \frac{1}{2} (l^a{}_{;a})^2 , \tag{1.6}$$

after proper normalization of the vector field. Such a congruence is said to be geodesic and shear-free. It emerged in the course of the next few years that these congruences played a particularly important part in the theory of exact solutions, and a new field of investigations opened up.

In this field, as in so many others, Alfred Schild made contributions of lasting significance. They began with the work of Pirani and Schild (1961, 1966) on the geometry of null congruences, laying bare the conformal structure of the subject. This work led to a generalization of the Goldberg-Sachs theorem. In empty space, as Goldberg and Sachs (1962) had shown, a shear-free geodesic congruence exists if and only if the Weyl tensor is algebraically degenerate in the sense that two or more of the Cartan directions coincide. It was Schild who perceived a curious asymmetry in this beautiful theorem: null shear-free congruences are conformally invariant objects; so is the Weyl tensor. But empty space is not. The result was a replacement of the empty space equations by a set of weaker conditions that were conformally invariant (Robinson and Schild, 1963).

Then came the Kerr-Schild metrics (1965), formed out of a flat background η_{ab} and a null vector field k_a, according to the equation

$$g_{ab} = \eta_{ab} + k_a k_b \,. \tag{1.7}$$

The use of a flat background metric was by no means new to Schild (1962), who had a long-standing interest in theories of the Whitehead type. The immediate stimulus for the investigation was given by Kerr's discovery of the metric that bears his name (Kerr, 1963). The work of Kerr and Schild produced a beautiful class of exact solutions and some fascinating problems. It seemed clear from the start that the Kerr-Schild metrics should admit some simple generalization, but nobody knew which direction this generalization should take.

In 1975 Plebanski and Schild returned to this problem, working on complex metrics. They looked at metrics of the form

$$g_{ab} = \eta_{ab} + k_{(a} l_{b)} \,, \tag{1.8}$$

where k_a and l_a are null, mutually orthogonal vectors. The basis of the double Kerr-Schild form is that in real four-dimensional spacetime, a symmetric tensor N_{ab} that is null,

$$N_{ab} N^{bc} = 0 \,, \tag{1.9}$$

must be proportional to $k_a k_b$, with k_a null. However, a generalization of the form of N_{ab} is possible in the complex domain, and there it is of the form $k_{(a} l_{b)}$. Plebanski and Schild draw attention in particular to the way in which the field equations simplify if the vector fields k_a and l_a are taken to be surface-forming. In that case the spacetime admits a congruence of totally null surfaces (Plebanski and Schild, 1976).

It seems appropriate to this occasion, therefore, for me to present a review of what is now known about congruences of null curves and totally null surfaces. I work throughout with complex metrics and vectors, but frequently on a real manifold.

I shall often be dealing with pairs of complex quantities that are in general completely independent but that have the property of reducing to complex conjugate pairs in the special case in which the metric and the null congruence are real. For this situation one requires a notation that is reminiscent of, but not identical to, the bar "‒" denoting complex conjugation. Following Robinson, Robinson, and Zund (1969), I use a tilde "~". In dealing with the geometry of real null geodesic congruences, for example, one is accustomed to combining the expansion θ and the rotation ω into $\rho = \theta + i\omega$. We then have $\bar{\rho} = \theta - i\omega$. Here, in complex congruences, θ and ω are complex; but we still need a symbol for $\theta - i\omega$, and we use $\tilde{\rho}$.

As far as symmetrization and antisymmetrization are concerned, I use the standard notation

$$T_{(ab)} := \frac{1}{2}(T_{ab} + T_{ba})$$
$$T_{[ab]} := \frac{1}{2}(T_{ab} - T_{ba})\,. \tag{1.10}$$

It seems curious that no standard notation for symmetrization followed by removal of the trace exists. I write

$$T_{\{ab\}} := \frac{1}{2}(T_{ab} + T_{ba}) - \frac{1}{4}T_c{}^c g_{ab}\,.$$

Some of the results that follow are the product of fairly long computations. I make no attempt to give these in full. I have tried, however, to provide sufficiently clear indications of the argument to turn verification of such results into manageable exercises for the reader.

2. EXPANSION AND ROTATION

Suppose that the vector field l^a is tangential to a congruence of null curves, so that

$$l^a l_a = 0\,, \qquad l^a l_{a;b} = 0\,. \tag{2.1}$$

The congruence determines l_a up to multiplication by an arbitrary function w:

$$l_a \to w l_a\,, \qquad l_{a;b} \to w l_{a;b} + l_a w_{,b}\,. \tag{2.2}$$

Thus if l^a is given at a point, its covariant derivative there is determined up to

$$l_{a;b} \rightarrow l_{a;b} + l_a w_b \,, \tag{2.3}$$

where w^b is an arbitrary vector. Our first problem is to construct fields algebraically from l_a and $l_{a;b}$ that are invariant under (2.3).

For this purpose, the tensor

$$L_{abc} := 2l_{[a}l_{b];c} \tag{2.4}$$

is useful: not only is it invariant under (2.3), but it determines $l_{a;b}$ up to this transformation. Using (2.1), we see that

$$l^a L_{abc} = 0 \,, \qquad l_{[a}L_{bc]d} = 0 \,. \tag{2.5}$$

As far as its first two indices are concerned, L^{abc} is a bivector (that is, an antisymmetric tensor of rank 2). As such, it splits into self-dual and anti-self-dual parts:

$$\begin{aligned} {}^+L_{abc} &= \frac{1}{2}(L_{abc} + i{}^*L_{abc}) \,, \\ {}^-L_{abc} &= \frac{1}{2}(L_{abc} - i{}^*L_{abc}) \,, \end{aligned} \tag{2.6}$$

with

$$i{}^*L_{abc} = \frac{1}{2}\eta_{abpq}L^{qp}{}_c \,, \qquad \eta_{abcd} := \sqrt{g}\; \varepsilon_{abcd} \,. \tag{2.7}$$

(In real spacetime the determinant g of the metric is negative, η_{abcd} is purely imaginary, and $^*L_{abc}$ is real. Note that in any case $** = -1$.) From (2.5):

$$l_a{}^+L^a{}_{bc} = 0 \,, \qquad l_a{}^-L^a{}_{bc} = 0 \,. \tag{2.8}$$

We use these tensors to define

$$m_a := {}^-L_{ab}{}^b \,, \qquad \tilde{m}_a := {}^+L_{ab}{}^b \,, \tag{2.9}$$

and

$$M_{ab} := {}^+L_{ap}{}^{q}\,{}^-L_{bq}{}^p \,. \tag{2.10}$$

The vectors m^a and $\tilde{m}^a$ can be constructed from the expansion,

$$\theta := \frac{1}{2}\, l^a{}_{;a} \,, \tag{2.11}$$

and the rotation,

$$\omega_{ab} := l_{[a,b]} \,, \tag{2.12}$$

only: writing

$$Q_{ab} := \theta g_{ab} - 2^{+}\omega_{ab}\,, \qquad \tilde{Q}_{ab} := \theta g_{ab} - 2^{-}\omega_{ab}\,, \tag{2.13}$$

we have

$$m_a = Q_{ab} l^b\,, \qquad \tilde{m}_a = \tilde{Q}_{ab} l^b\,. \tag{2.14}$$

Furthermore, m_a and $\tilde{m}_a$ determine the expansion and the rotation up to the transformation induced by (2.3). They are known as the *left expansion* and the *right expansion*, respectively. We define also the *compound expansion*,

$$n_a := Q_{ab} \tilde{Q}^{bc} l_c\,. \tag{2.15}$$

Two rules of bivector algebra are needed here, both concerned with matrix multiplication (that is, with multiplication followed by contraction over adjacent indices): a self-dual bivector and an anti-self-dual bivector commute; and the product of a pair of self-dual bivectors or a pair of anti-self-dual bivectors is the sum of a pure trace and a bivector of the original chirality.

By using these rules, we can easily calculate that

$$l^a n_a + m^a \tilde{m}_a = 0 \tag{2.16}$$

and that all the other scalar products of the four vectors vanish identically. Thus

$$\alpha g_{ab} = 2 l_{(a} n_{b)} - 2 m_{(a} \tilde{m}_{b)}\,, \tag{2.17}$$

for some function α.

To investigate the orientation of this tetrad (l^a, n^a, m^a, $\tilde{m}^a$), we invoke a third rule of bivector algebra: the skew-symmetrized product of a bivector in a pure chirality state with a trace-free symmetrical tensor of the second rank is a bivector of the opposite chirality. Thus

$$\begin{aligned} {}^{+}(\alpha_{[i\,|\,p\,|}\,\omega^{p}{}_{j]}) &= \alpha_{[i\,|\,p\,|}\,{}^{-}\omega^{p}{}_{j]}\,, \\ {}^{-}(\alpha_{[i\,|\,p\,|}\,\omega^{p}{}_{j]}) &= \alpha_{[i\,|\,p\,|}\,{}^{+}\omega^{p}{}_{j]}\,, \end{aligned} \tag{2.18}$$

for any trace-free symmetrical α_{ij}. In the special case

$$\alpha_{ij} = l_i l_j\,, \tag{2.19}$$

we get

$$m_{[a} l_{b]} = {}^{-}(l_{[a} \dot{l}_{b]})\,, \qquad \tilde{m}_{[a} l_{b]} = {}^{+}(l_{[a} \dot{l}_{b]})\,, \tag{2.20}$$

where

$$\dot{l}_a := l_{a;b} l^b = 2\theta l_a - m_a - \tilde{m}_a\,. \tag{2.21}$$

If α is nowhere zero, we obtain $\alpha = 1$ from a transformation (2.2). We then have

$$g_{ab} = 2l_{(a}n_{b)} - 2m_{(a}\tilde{m}_{b)}\,, \tag{2.22}$$

$$\eta^{abcd}l_a m_b \tilde{m}_c n_d = 1\,, \tag{2.23}$$

$$m^a l^b l_{a;b} = \tilde{m}^a l^b l_{a;b} = 1\,, \tag{2.24}$$

on account of (2.17), (2.20), and (2.21). Returning to the definition of the expansion vectors, we find that

$$m^a \tilde{m}^b l_{a;b} = \tilde{m}^a m^b l_{a;b} = 0\,. \tag{2.25}$$

These conditions, together with the fact that l^a is tangential to the congruence, determine the tetrad up to the transformations generated by

$$(l_a,\ m_a,\ \tilde{m}_a,\ n_a) \rightarrow (il_a,\ -m_a,\ -\tilde{m}_a,\ in_a)\,. \tag{2.26}$$

In the real case, we can fix the tetrad completely by requiring that l_a should be real and future-directed. Whether this tetrad is useful in practice has yet to be determined: in this paper, we are chiefly concerned with the case

$$\alpha = 0\,, \tag{2.27}$$

in which it is not available.

3. THE ASSOCIATED BIVECTORS

In tangent space, l^a picks out two totally null planes (i.e., two-dimensional subspaces in which every vector is null), and these determine a pair of bivectors up to

$$F_{ab} \rightarrow \Omega F_{ab}\,, \qquad \tilde{F}_{ab} \rightarrow \tilde{\Omega}\tilde{F}_{ab}\,. \tag{3.1}$$

The defining equations are

$$F_{ab} = i{}^*F_{ab}\,, \qquad F_{ab}l^b = 0\,, \tag{3.2}$$

and

$$\tilde{F}_{ab} = -i{}^*\tilde{F}_{ab}\,, \qquad \tilde{F}_{ab}l^b = 0\,. \tag{3.3}$$

It follows from (2.8) that there exist vectors λ_a and $\tilde{\lambda}_a$, with the transformation

$$\lambda_a \rightarrow \tilde{\Omega}^{-1}\lambda_a\,, \qquad \tilde{\lambda}_a \rightarrow \Omega^{-1}\tilde{\lambda}_a \tag{3.4}$$

under (3.1), such that

$$L_{abc} = \tilde{F}_{ab}\lambda_c + F_{ab}\tilde{\lambda}_c\,. \tag{3.5}$$

The vectors of one-sided expansion may be written as

$$m_a = \tilde{F}_{ab}\lambda^b\,, \qquad \tilde{m}_a = F_{ab}\tilde{\lambda}^b\,. \tag{3.6}$$

The derivatives of the auxiliary bivectors along the congruence,

$$\dot{F}_{ab} := F_{ab;c}l^c , \qquad \dot{\tilde{F}}_{ab} := \tilde{F}_{ab;c}l^c , \tag{3.7}$$

are self-dual and anti-self-dual, respectively. To investigate them, we introduce the bivectors

$$U_{ab} := Q_{[a\,|\,r\,|}\,F^r{}_{b]} , \qquad \tilde{U}_{ab} := \tilde{Q}_{[a\,|\,r\,|}\,\tilde{F}^r{}_{b]} . \tag{3.8}$$

Using our second rule of bivector algebra, we see that

$$U_{ab} = i^*U_{ab} , \qquad \tilde{U}_{ab} = -i^*\tilde{U}_{ab} . \tag{3.9}$$

In writing out U^{ab} explicitly, we can replace ${}^+\omega^{ab}$ by ω^{ab}: because of the first rule, the anti-self-dual part makes no contribution. We obtain

$$U_{ab} = F_a{}^p l_{[p,b]} - l_{[a,p]}F^p{}_b + \frac{1}{2}\, l^p{}_{;p}F_{ab} , \tag{3.10}$$

and conclude that

$$(\dot{F}_{ab} + 2U_{ab})l^b = 0 . \tag{3.11}$$

The bivector $\dot{F}^{ab} + 2U^{ab}$ is thus a solution of (3.2), and there exists, in consequence, a scalar ξ such that

$$\dot{F}_{ab} + 2U_{ab} = \xi F_{ab} . \tag{3.12}$$

Similarly, the anti-self-dual field is propagated according to the equation

$$\dot{\tilde{F}}_{ab} + 2\tilde{U}_{ab} = \tilde{\xi}\tilde{F}_{ab} , \tag{3.13}$$

for some $\tilde{\xi}$. Under the transformation (2.2),

$$\xi \rightarrow w\xi , \qquad \tilde{\xi} \rightarrow w\tilde{\xi} ; \tag{3.14}$$

under the transformation (3.1),

$$\xi \rightarrow \xi + (ln\Omega)_{,p}l^p , \qquad \tilde{\xi} \rightarrow \tilde{\xi} + (ln\tilde{\Omega})_{,p}l^p . \tag{3.15}$$

We can recover our tetrad from these bivectors by considering their *Maxwell products*: that is, objects of the form

$$T_{ab} := X_{ap}Y^p{}_b , \tag{3.16}$$

where

$$X_{ab} = i^*X_{ab} , \qquad Y_{ab} = -i^*Y_{ab} . \tag{3.17}$$

With a generic Maxwell product there are associated four null vectors, the common eigenvectors of the factors. If one of the bivectors is null, the vectors come together in two pairs; if both, in a quadruplet. In all cases, we can write the Maxwell product as

$$T_{ab} = p_{\{a}q_{b\}}, \tag{3.18}$$

with

$$p_a p^a = p_a q^a + 4\xi\eta = q_a q^a = 0, \tag{3.19}$$

where p^a, q^a are eigenvectors of X_{ab} and Y_{ab}:

$$\begin{aligned} X_{ab}p^b &= \xi p_a, & Y_{ab}q^b &= \eta q_a, \\ X_{ab}q^b &= -\xi q_a, & Y_{ab}p^b &= -\eta p_a. \end{aligned} \tag{3.20}$$

Unless $\xi\eta = 0$, (3.18) follows from (3.19) and (3.20): in the exceptional case, a further restriction is needed to fix the scale.

From a pair of self-dual bivectors and a pair of anti-self-dual bivectors, we can form four Maxwell products. Any three of them are sufficient to determine the fourth: this is a consequence of the *unscrambling identity*,

$$2T_{a[d}g_{c]b} + 2T_{b[c}g_{d]a} = X_{ab}Y_{cd} + Y_{ab}X_{cd}. \tag{3.21}$$

We also conclude from it that a Maxwell product vanishes only when at least one of its factors is zero.

Applying these general considerations to the present case, we obtain immediately

$$F_{ab}\tilde{F}^b{}_c = f l_a l_c, \tag{3.22}$$

for some nonzero f. Using this result and the basic rules of bivector algebra, we obtain from (3.8)

$$U_{ab}\tilde{F}^b{}_c = f l_{(a}m_{c)}, \qquad \tilde{U}_{ab}F^b{}_c = f l_{(a}\tilde{m}_{c)}, \tag{3.23}$$

and

$$U_{ab}\tilde{U}^b{}_c = f l_{\{a}n_{c\}}. \tag{3.24}$$

It follows that there exist scalars τ and $\tilde{\tau}$ such that

$$U_{ab}l^b = \tau l_a, \qquad \tilde{U}_{ab}l^b = \tilde{\tau} l_a, \tag{3.25}$$

and

$$4\tau\tilde{\tau} = f\alpha. \tag{3.26}$$

These are known as the *left curvature* and the *right curvature*, respectively. Under the transformation (3.1),

$$\tau \to \Omega\tau, \qquad \tilde{\tau} \to \tilde{\Omega}\tilde{\tau}. \tag{3.27}$$

For some purposes, it is desirable to determine the auxiliary bivectors more precisely. An obvious restriction is

$$f = 1. \tag{3.28}$$

If $\alpha \neq 0$, we may also require that

$$\tau = \tilde{\tau}\,. \tag{3.29}$$

The functions in (3.1) are thereby restricted to

$$\Omega = \tilde{\Omega} = \pm 1\,. \tag{3.30}$$

4. SHEAR

The first covariant derivative of the tangent vector decomposes into rotation ω_{ab}, expansion θ, and shear σ_{ab}:

$$l_{a;b} = \omega_{ab} + \frac{1}{2}\theta g_{ab} + \sigma_{ab}\,, \tag{4.1}$$

with

$$\sigma_{ab} := l_{\{a;b\}}\,. \tag{4.2}$$

They are not entirely independent: because l_a is null,

$$2\sigma_{ab}l^b = \theta l_a - m_a - \tilde{m}_a\,. \tag{4.3}$$

We have used rotation and expansion in constructing the expansion vectors. We shall now see what can be done with shear.

Our object is to construct a tetrad. For this purpose, we define a pair of bivectors:

$$V_{ab} := -2F_{[a\,|\,p\,|}\,\sigma^p{}_{b]}\,, \qquad \tilde{V}_{ab} := -2\tilde{F}_{[a\,|\,p\,|}\,\sigma^p{}_{b]}\,. \tag{4.4}$$

We then have

$$l^a(U_{ab} + V_{ab}) = 0 = l^a(\tilde{U}_{ab} + \tilde{V}_{ab}) \tag{4.5}$$

immediately, and hence

$$V_{ab}l^b = -\tau l_a\,, \qquad \tilde{V}_{ab}l^b = -\tilde{\tau} l_a\,, \tag{4.6}$$

from (3.25). Moreover,

$$V_{ab} = -i{*}V_{ab}\,, \qquad \tilde{V}_{ab} = i{*}\tilde{V}_{ab}\,, \tag{4.7}$$

from (3.2), (3.3), and the third rule of bivector algebra. It follows that

$$V_{ab}F^b{}_c = f l_{(a}\mu_{c)}\,, \qquad \tilde{V}_{ab}F^b{}_c = f l_{(a}\tilde{\mu}_{c)}\,, \tag{4.8}$$

and

$$V_{ab}\tilde{V}^b{}_c = f l_{\{a}\nu_{c\}} \tag{4.9}$$

for some vectors μ_a, $\tilde{\mu}_a$, and ν_a, subject to

$$\alpha g_{ab} = 2l_{(a}\nu_{b)} - 2\mu_{(a}\tilde{\mu}_{b)} \,. \tag{4.10}$$

These are the vectors of *shear*: the *left shear* (μ_a), the *right shear* ($\tilde{\mu}_a$), and the *compound shear* (ν_a). The tetrad l_a, μ_a, $\tilde{\mu}_a$, ν_a has the opposite orientation to l_a, m_a, $\tilde{m}_a$, n_a:

$$\mu_{[a}l_{b]} = i{}^*(\mu_{[a}l_{b]}) \,, \qquad \tilde{\mu}_{[a}l_{b]} = -i{}^*(\tilde{\mu}_{[a}l_{b]}) \,. \tag{4.11}$$

All this follows directly from (4.6) and (4.7). More detailed calculation shows that

$$\mu_a = F_{ab}\lambda^b \,, \qquad \tilde{\mu}_a = \tilde{F}_{ab}\tilde{\lambda}^b \,, \tag{4.12}$$

and

$$\nu_a = \left(\frac{1}{2}\sigma_{pq}\sigma^{qp}\delta_a^b - 2\sigma_{aq}\sigma^{qb} \right) l_b \,. \tag{4.13}$$

Under (2.2), the vectors of shear transform in the same way as their counterparts in the expansion triad:

$$\mu_a \to w^2\mu_a \,, \qquad \tilde{\mu}_a \to w^2\tilde{\mu}_a \,, \qquad \nu_a \to w^3\nu_a \,. \tag{4.14}$$

Under the transformation (3.1), however,

$$\mu_a \to (\Omega/\tilde{\Omega})\mu_a \,, \qquad \tilde{\mu}_a \to (\tilde{\Omega}/\Omega)\tilde{\mu}_a \,. \tag{4.15}$$

It is thus impossible to construct the vectors of one-sided shear from L_{abc} alone without the use of F_{ab} and $\tilde{F}_{ab}$. Their product, however, is given by

$$\mu_a\tilde{\mu}_b = M_{ab} \,, \tag{4.16}$$

with M_{ab} defined in (2.10).

In the case $\alpha \neq 0$, we can eliminate the transformation (4.15) by requiring that (3.29) should be satisfied. The tetrads of expansion and shear then differ by a null rotation:

$$(l_a,\ \tilde{\mu}_a,\ \mu_a,\ \nu_a) \to N_a{}^b(l_b,\ m_b,\ \tilde{m}_b,\ n_b) \,, \tag{4.17}$$

with

$$\alpha^3 N_{ab} := \alpha^3 g_{ab} - 2\alpha(\beta\tilde{m}_{[a} + \tilde{\beta}m_{[a})l_{b]} - \beta\tilde{\beta}l_al_b \,, \tag{4.18}$$

$$\beta := m_a\nu^a \,, \qquad \tilde{\beta} = \tilde{m}_a\nu^a \,. \tag{4.19}$$

We note that

$$\mu^al^bl_{a;b} = \tilde{\mu}^al^bl_{a;b} = 1 \tag{4.20}$$

$$\mu^a\mu^bl_{a;b} = \tilde{\mu}^a\tilde{\mu}^bl_{a;b} = 0 \,. \tag{4.21}$$

5. SEMI-GEODESICS

Let us consider a single curve, not necessarily null. The condition for the curve to be geodesic is that a certain bivector should vanish, namely

$$L_{ab} := l_{[a}\dot{l}_{b]}\,, \tag{5.1}$$

where l^a is the tangent vector and $\dot{l}^a$ its covariant derivative along the curve. The bivector decomposes into self-dual and anti-self-dual parts: the geodesic equation decomposes accordingly into

$$^{+}L_{ab} = 0 \tag{5.2}$$

and

$$^{-}L_{ab} = 0\,. \tag{5.3}$$

We say that a curve is *right-geodesic* if it satisfies the first equation, *left-geodesic* if it satisfies the second, and *semi-geodesic* if either equation holds. In complexified Minkowski space, the curve

$$x = ct\,, \qquad y = iz = f(t)\,, \tag{5.4}$$

for example, is semi-geodesic for any f, but geodesic only when $f'' = 0$.

Semi-geodesics are, of course, objects of complex geometry: in the real case, (5.2) and (5.3) are equivalent. There is further restriction, embodied in the identities

$$\begin{aligned}(l_p l^p)\dot{l}_a\dot{l}_b - 2(l_p\dot{l}^p)l_{(a}\dot{l}_{b)} + (\dot{l}_p\dot{l}^p)l_a l_b &= 8L_{ap}{}^{+}L_b{}^p \\ &= 8L_{ap}{}^{-}L_b{}^p\,.\end{aligned} \tag{5.5}$$

If one and only one of the equations (5.2) and (5.3) holds, then l_a and $\dot{l}_a$ are linearly independent, and the left-hand side of (5.5) vanishes. Consequently,

$$l_p l^p = 0\,, \qquad l_p\dot{l}^p = 0\,, \qquad \dot{l}_p\dot{l}^p = 0\,, \tag{5.6}$$

and the curve is null. One can understand the result in geometrical terms as follows. At any point where the original bivector is nonzero, it picks out a two-dimensional subspace of tangent space: that spanned by l_a and $\dot{l}_a$. Its dual determines similarly the orthogonal subspace. In a semi-geodesic, the two bivectors are proportional, and their subspaces coincide: hence (5.6).

Suppose now that we have a congruence of null curves. From (2.20) it follows that the congruence is left-geodesic if and only if

$$m_a = \rho l_a \tag{5.7}$$

for some ρ. Comparing equations (3.22) and (3.23), we see that this is equivalent to

$$U_{ab} = \rho F_{ab} \,. \tag{5.8}$$

Writing $\lambda := 2\rho - \xi$, we now get

$$F_{ab;c} l^c + \lambda F_{ab} = 0 \tag{5.9}$$

from (3.12). From (3.9) and (3.25), it follows that a congruence satisfies the left-geodesic condition (5.8) if and only if its left curvature vanishes:

$$\tau = 0 \,. \tag{5.10}$$

This, because of (4.6) and (4.7), is necessary and sufficient for a scalar σ to exist such that

$$V_{ab} = \sigma \tilde{F}_{ab} \,. \tag{5.11}$$

Turning to (4.8), we see that the last equation is equivalent to

$$\mu_a = \sigma l_a \,. \tag{5.12}$$

This too, therefore, is a necessary and sufficient condition for the congruence to be left-geodesic.

In a left-geodesic congruence, $\alpha = 0$ from (3.26) and (5.10);

$$n_a = \rho \tilde{m}_a \tag{5.13}$$

from (2.17) and (5.7); and

$$\nu_a = \sigma \tilde{\mu}_a \tag{5.14}$$

from (4.10) and (5.12). Each of the tetrads, therefore, collapses into two pairs of parallel vectors:

$$(l_a,\ m_a,\ \tilde{m}_a,\ n_a) \rightarrow (l_a,\ \rho l_a,\ \tilde{m}_a,\ \rho \tilde{m}_a)\,,$$

$$(l_a,\ \mu_a,\ \tilde{\mu}_a,\ \nu_a) \rightarrow (l_a,\ \sigma l_a,\ \tilde{\mu}_a,\ \sigma \tilde{\mu}_a)\,.$$

A special case of considerable importance is the fully geodesic congruence. There all vectors in the two tetrads are parallel to l_a; and there are two additional scalars, $\tilde{\rho}$ and $\tilde{\sigma}$. Under the transformation (2.2),

$$(\rho,\ \tilde{\rho},\ \sigma,\ \tilde{\sigma}) \rightarrow w(\rho,\ \tilde{\rho},\ \sigma,\ \tilde{\sigma})\,; \tag{5.15}$$

under (3.1),

$$(\rho,\ \tilde{\rho},\ \sigma,\ \tilde{\sigma}) \rightarrow (\rho,\ \tilde{\rho},\ \sigma\Omega/\tilde{\Omega},\ \tilde{\sigma}\tilde{\Omega}/\Omega)\,. \tag{5.16}$$

On account of (5.9) and its anti-self-dual counterpart, we may take

$$\dot{F}_{ab} = 0 = \dot{\tilde{F}}_{ab} \,. \tag{5.17}$$

If we combine this with (3.22) and the normalization (3.28), we get

$$\dot{l}_a = 0\,, \tag{5.18}$$

so that the geodesics are affinely normalized. Then

$$\rho = \theta + i\omega\,, \qquad \bar{\rho} = \theta - i\omega\,, \tag{5.19}$$

where θ is given by (2.11), and

$$\omega^2 = \frac{1}{2}\, l_{[a;b]} l^{a;b}\,. \tag{5.20}$$

For the compound shear, we have

$$\nu_a = \sigma\tilde{\sigma} l_a\,, \qquad \sigma\tilde{\sigma} = \frac{1}{2}\, l_{(a;b)} l^{a;b} - \theta^2\,. \tag{5.21}$$

The transformations (5.15) and (5.16) remain, but the coefficients are constant on each geodesic. The scalars ρ and σ are now the complex expansion and complex shear of R. K. Sachs (1962).

Returning to the semi-geodesic case, we say that a congruence satisfying

$$m_a = 0 \tag{5.22}$$

is *expansion-free on the left*; if

$$\mu_a = 0\,, \tag{5.23}$$

it is *shear-free on the left*. There is a basic difference between these two conditions, as we shall see: the first is a restriction on the congruence; the second is a restriction on an equivalence class of congruences.

6. LEFT-EQUIVALENT CONGRUENCES

Suppose that k^a and l^a are tangential to congruences of null curves. The congruences coincide if

$$^{+}(k_{[a} l_{b]}) = 0 \tag{6.1}$$

and

$$^{-}(k_{[a} l_{b]}) = 0\,. \tag{6.2}$$

We say that they are *right-equivalent* if (6.1) holds and *left-equivalent* if (6.2) holds. In either case, k_a and l_a lie in a null plane. From (3.2),

$$-(k_{[a} l_{p]}) F^p{}_b = \frac{1}{2}\, l_{(a} F_{b)p} k^p \tag{6.3}$$

identically; (6.2), therefore, is equivalent to

$$F_{ab} k^b = 0\,; \tag{6.4}$$

and a pair of null congruences are left-equivalent if and only if they belong to the same self-dual null bivector.

To investigate this bivector, we introduce two vectors:

$$J^a := F^{ab}{}_{;b}\,, \qquad \Sigma_a := F_{ab}J^b\,. \tag{6.5}$$

Because F_{ab} is null and self-dual,

$$F_{ab}\Sigma^b = 0\,, \tag{6.6}$$

and under (3.1),

$$\Sigma_a \to \Omega^2\Sigma_a\,. \tag{6.7}$$

In any region where the *distortion vector* Σ^a does not vanish, it picks out a single null congruence from the left-equivalence class defined by F_{ab}.

For any congruence of this class, a tangent vector l_a satisfies

$$F_{bc}l_a + F_{ca}l_b + F_{ab}l_c = 0\,, \tag{6.8}$$

on account of (3.2). Taking the covariant divergence, we obtain

$$2J_{[a}l_{b]} = \dot{F}_{ab} + F_{ab}l^c{}_{;c} + F_b{}^c l_{a;c} + F^c{}_a l_{b;c}\,; \tag{6.9}$$

and hence, from (3.7), (3.9), and (4.4),

$$2J_{[a}l_{b]} = \xi F_{ab} - U_{ab} + V_{ab}\,. \tag{6.10}$$

Hence,

$$2^-(J_{[a}l_{b]})F^b{}_c = V_{ab}F^b{}_c\,, \tag{6.11}$$

from which, using (4.8) and the identity (6.3), we obtain

$$\Sigma_a = f\mu_a\,. \tag{6.12}$$

Up to a disposable factor, therefore, the left shear is the same for all congruences of the left-equivalence class. We know, moreover, that a congruence is left-geodesic if and only if the left shear is proportional to the tangent vector. This condition can now be written as

$$l_{[a}\Sigma_{b]} = 0\,. \tag{6.13}$$

If the distortion vector does not vanish, it is tangential to the one left-geodesic congruence in the equivalence class; if it does vanish, all the congruences are shear-free on the left.

In either case, if l_a is left-geodesic, then (6.10) reduces to

$$2J_{[a}l_{b]} = (\xi - \rho)F_{ab} + \sigma\tilde{F}_{ab} \tag{6.14}$$

on account of (5.8) and (5.11). For a given left-geodesic congruence, we can always make

$$2J_{[a}l_{b]} = \sigma\tilde{F}_{ab}\,, \tag{6.15}$$

which is a simple matter of using the transformation (3.1) to make $\xi = \rho$. It then follows from (3.12) and (5.8) that

$$F_{ab;c}l^c + \rho F_{ab} = 0\,, \tag{6.16}$$

where ρ is the scalar of left expansion, defined by

$$2^+(l_{[a,b]})l^b + \left(\rho - \frac{1}{2}l^b{}_{;b}\right) l_a = 0\,. \tag{6.17}$$

What happens if the distortion vector vanishes, and every congruence in the left-equivalence class is semi-geodesic? From (3.2),

$$F_{a[b}F_{cd]} = 0\,, \tag{6.18}$$

identically, and

$$F_{a[b}F_{cd,e]} = 0\,, \tag{6.19}$$

if and only if

$$\Sigma_a = 0\,. \tag{6.20}$$

From the integration theorem of Frobenius, however, (6.18) and (6.19) are necessary and sufficient conditions for the (local) existence of scalars t, u, v, such that

$$F_{ab} = 2tu_{,[a}v_{,b]}: \tag{6.21}$$

F^{ab} is surface-forming, therefore, if and only if the distortion vanishes. In that case, a transformation of the form

$$(t, u, v) \rightarrow (\Omega t, u, v) \tag{6.22}$$

combined with (3.1) gives

$$t = t(u, v)\,, \tag{6.23}$$

which is equivalent to

$$F_{[ab;c]} = 0\,, \tag{6.24}$$

or, inasmuch as F_{ab} is self-dual, to

$$J_a = 0\,; \tag{6.25}$$

and since $\sigma = 0$ because of (6.20), equation (6.16) is satisfied for every congruence of curves associated with F_{ab}.

The scalars u and v are determined up to

$$u \rightarrow \hat{u}(u, v)\,, \qquad v \rightarrow \hat{v}(u, v)\,, \qquad \partial(\hat{u}, \hat{v})/\partial(u, v) \neq 0\,, \tag{6.26}$$

by (6.21). Applying here the arguments that led to (5.6), we see that

$$g^{ab}u_{,a}u_{,b} = g^{ab}u_{,a}v_{,b} = g^{ab}v_{,a}v_{,b} = 0\,. \tag{6.27}$$

For any $l(u, v)$, moreover, $l_{,a}$ is shear-free on the left. From the second of equations (5.21), therefore,

$$u_{;p}{}^{q}u_{;q}{}^{p} = \frac{1}{2}\,(u_{;p}{}^{p})^{2}\,, \tag{6.28}$$

and

$$u_{;p}{}^{q}v_{;q}{}^{p} = \frac{1}{2}\,u_{;p}{}^{p}v_{;q}{}^{q}\,. \tag{6.29}$$

The surfaces of constant u and v are invariant under (6.26) and as we see from (6.27), totally null. We distinguish between *left-handed* null surfaces, like these, in which the surface element is self-dual, and *right-handed* null surfaces in which it is anti-self-dual. A congruence of null curves is shear-free on both sides if and only if it is the intersection of two congruences of totally null surfaces, one left-handed and the other right-handed.

With any real congruence of this kind, we can associate a null solution of Maxwell's equation for empty space: $F_{ab} + \tilde{F}_{ab}$. Then (6.16) becomes the equation of Lichnerowicz (1962) for the propagation of pure electromagnetic radiation.

7. CONFORMAL TRANSFORMATION

Yet another difference between the tetrads of shear and expansion emerges when we consider the transformation

$$g_{ab} \rightarrow \omega g_{ab}\,, \qquad \eta_{abcd} \rightarrow \omega^{2}\eta_{abcd}\,, \tag{7.1}$$

with arbitrary nonzero ω. It is convenient to assume that

$$l_{a} \rightarrow \omega l_{a}\,, \qquad F_{ab} \rightarrow F_{ab}\,, \qquad \tilde{F}_{ab} \rightarrow \tilde{F}_{ab}: \tag{7.2}$$

this is a restriction of no great importance, because we still have transformations (2.2) and (3.1) at our disposal. We find that

$$l_{a;b} \rightarrow \omega l_{a;b} + l_{[a}\omega_{,b]} - \frac{1}{2}\,l^{p}\omega_{,p}g_{ab}\,, \tag{7.3}$$

$$\dot{l}_{a} \rightarrow \omega \dot{l}_{a}\,, \tag{7.4}$$

and hence

$${}^{+}L^{ab} \rightarrow {}^{+}L^{ab}\,, \qquad {}^{-}L^{ab} \rightarrow {}^{-}L^{ab}: \tag{7.5}$$

thus right-geodesics and left-geodesics are conformally invariant. Using (7.3), we see that

$$l^a,\ \mu^a,\ \tilde{\mu}^a,\ \nu^a \tag{7.6}$$

and

$$m^a - \tilde{m}^a \tag{7.7}$$

are all invariant under (7.1) and (7.2). On the other hand,

$$\theta \rightarrow \theta - \omega^{-1}\omega_{,p}l^p ; \tag{7.8}$$

hence,

$$m^a + \tilde{m}^a \rightarrow m^a + \tilde{m}^a + 2\chi l^a \tag{7.9}$$

from (2.21), and

$$n^a \rightarrow n^a + \chi(m^a + \tilde{m}^a) + \chi^2 l^a \tag{7.10}$$

from (2.17), where

$$\chi := -\omega^{-1}\omega_{,p}l^p . \tag{7.11}$$

These results were first obtained by Pirani and Schild (1961, 1966) for fully geodesic congruences.

8. THE DEVIATION VECTOR

The self-dual bivector F^{ab} made its appearance as a convenient device for describing null vectors, but it has now become the primary object of our investigation. Its most obvious feature is that it is null as well as self-dual: from (3.2) and the second rule of bivector algebra,

$$F_{ab}F^{bc} = 0 . \tag{8.1}$$

This makes it possible for us to define an important vector field. Writing

$$F'_{ab} := F_{ap}F^p{}_{b;c}A^c , \tag{8.2}$$

where A^c is arbitrary, we have

$$F'_{(ab)} = 0 , \qquad F'_{ab}F^{bc} = 0 , \tag{8.3}$$

from which it follows that F'^{ab} is proportional to F^{ab}. Thus there exists a vector Θ_c such that

$$F_{ap}F^p{}_{b;c} = F_{ab}\Theta_c . \tag{8.4}$$

This we call the *deviation vector*. We define also

$$\Phi_{ab} := \Theta^c{}_{;c}F_{ab} - F_{ab;c}\Theta^c . \tag{8.5}$$

Under the transformation (3.1),

$$\Theta_a \to \Omega\Theta_a, \qquad \Phi_{ab} \to \Omega^2\Phi_{ab}. \tag{8.6}$$

The distortion vector is the projection of the deviation into the plane of F_{ab}: contracting (8.4), we have

$$\Sigma_a = F_{ab}\Theta^b. \tag{8.7}$$

Since F_{ab} is self-dual,

$$F^{cd} = \frac{1}{2}\eta^{cdpq}F_{qp}; \tag{8.8}$$

hence,

$$F_{ab;c} + F_{ca;b} + F_{bc;a} = \eta_{abcr}J^r \tag{8.9}$$

from the definition of J^a, and

$$F_{ab}\Theta^d + F^d{}_a\Theta_b + F_b{}^d\Theta_a = 2i{*}(\Sigma_{[a}\delta^d_{b]}) \tag{8.10}$$

from (8.7). Considering that

$$F^a{}_p{}^+X^{pb} = F^{[a}{}_pX^{|p|b]} + \frac{1}{4}F_{pq}X^{qp}g^{ab}, \tag{8.11}$$

for any bivector X^{ab}, we get

$$F^a{}_p\{F^{pb}{}_{;c} - 4^+(\delta^{[p}_c\Theta^{b]})\} = g^{ab}\Sigma_c - 4^+(\delta^{[a}_c\Sigma^{b]}) \tag{8.12}$$

from (8.4), (8.7), and (8.10). It follows that the distortion vector vanishes if and only if (8.3) is satisfied by

$$F'^{ab} = \{F^{ab}{}_{;c} - 4^+(\delta^{[a}_c\Theta^{b]})\}A^c \tag{8.13}$$

for every A^c. A necessary and sufficient condition for vanishing distortion, therefore, is that

$$F^{ab}{}_{;c} = F^{ab}\Psi_c + 4^+(\delta^{[a}_c\Theta^{b]}) \tag{8.14}$$

for some Ψ_c. Under (3.1),

$$\Psi_c \to \Psi_c + (ln\Omega)_{,c}, \tag{8.15}$$

and $\Psi_{[a,b]}$ is invariant. The trace of (8.14) is

$$J^a = F^{ab}\Psi_b - 3\Theta^a; \tag{8.16}$$

its trace-free part is equivalent to Sommers' equation in spinors (Sommers, 1976):

$$\nabla_{A'(A}\kappa_{B)} = \xi_{A'(A}\kappa_{B)}$$

with

$$F_{ab} \leftrightarrow \kappa_A \kappa_B \varepsilon_{A'B'}, \qquad \Psi_a \leftrightarrow 2\xi_{AA'}.$$

In the special case of vanishing deviation, (8.14) reduces to Finley's equation (J. D. Finley, private communication, 1977),

$$F_{ab;c} = F_{ab}\Psi_c, \tag{8.17}$$

and the bivector is recurrent.

We shall obtain another useful identity for the deviation vector. Contracting (8.8) with (8.9), adding (8.10), and using (8.4), we get

$$F_{ab;c}F^{cd} + F_{ab}(\Theta^d + J^d) = 4\,{}^+(\Sigma_{[a}\delta^d_{b]}). \tag{8.18}$$

If the distortion vector vanishes, this reduces to

$$F_{ab;c}F^{cd} + F_{ab}(\Theta^d + J^d) = 0. \tag{8.19}$$

Suppose, conversely, that (8.19) holds for some vector Θ^d. Contracting with F_{pd}, we have

$$F_{pq}(\Theta^q + J^q) = 0 \tag{8.20}$$

from (8.1). Let k^a and l^a be any independent vectors in the plane of F_{ab}, normalized so that

$$F^{cd} = 2k^{[c}l^{d]}. \tag{8.21}$$

Because of (8.20), there exist scalars, κ and λ, such that

$$\Theta^d + J^d = \kappa l^d - \lambda k^d. \tag{8.22}$$

Substituting into (8.19), we obtain (5.9) and the corresponding equation for propagation along k_a:

$$F_{ab;c}k^c + \kappa F_{ab} = 0. \tag{8.23}$$

Thus k^a and l^a are both left-geodesic. From this, it follows that F_{ab} is surface-forming.

The existence of a vector Θ^c subject to (8.19) is thus a necessary and sufficient condition for the distortion vector to vanish.

From now on, we shall suppose that this condition is satisfied—a very important restriction. From (8.22), (8.23), and (5.9)

$$F_{ab;c}(\Theta^c + J^c) = 0. \tag{8.24}$$

From (8.7), Θ^a is left-geodesic; consequently,

$$\Phi_{ab} = \Phi F_{ab}, \tag{8.25}$$

for some Φ, and hence

$$(\Phi - \Theta^c{}_{;c})F_{ab} = F_{ab;c}J^c \tag{8.26}$$

from the definition (8.5). One might think of Φ as something like the divergence of the expansion vector, inasmuch as

$$J_a = 0 \Rightarrow \Phi = \Theta^c{}_{;c}\,. \tag{8.27}$$

It has, however, a simple transformation,

$$\Phi \rightarrow \Omega\Phi\,, \tag{8.28}$$

under (3.1).

When $J_a = 0$, (8.22) reduces to

$$\Theta^d = \rho' l^d - \rho k^d\,, \tag{8.29}$$

where ρ' and ρ are the left-expansion scalars of k_a and l_a, respectively. Hence, using (8.21),

$$2l_{[a}\Theta_{b]} = \rho F_{ab}: \tag{8.30}$$

a result, incidentally, that is invariant under the transformation (3.1) and therefore not dependent on the vanishing of J_a. Suppose that Θ_a is not zero. It then follows from (8.30) that there is one and only one congruence of curves belonging to F_{ab} for which the left expansion vanishes—the congruence tangential to the deviation vector. If, however, the deviation vector vanishes, then all congruences in the equivalence class are expansion-free on the left. In either case, because the left expansion of Θ_a vanishes,

$$\Theta_{a;b}\Theta^b = \Theta^b{}_{;b}\Theta_a - \Omega_a\,, \tag{8.31}$$

$$2{*}(\Theta_{[a,b]})\Theta^b = -i\Omega_a\,, \tag{8.32}$$

where Ω^a is the right expansion of Θ^a. This is a vector constructed from second covariant derivatives of F_{ab}. Before examining it further, we must look at the curvature tensor.

9. A RIGHT-EXPANSION VECTOR

The Riemann tensor decomposes into self-dual and anti-self-dual parts on each antisymmetrical pair of indices:

$$R_{abcd} = {}^{+}R^{+}_{abcd} + {}^{+}R^{-}_{abcd} + {}^{-}R^{+}_{abcd} + {}^{-}R^{-}_{abcd} \tag{9.1}$$

and

$${}^{+}R^{+}_{abcd} + {}^{-}R^{-}_{abcd} = C_{abcd} + \frac{1}{6}Rg_{a[d}g_{c]b}\,, \tag{9.2}$$

$$^{-}R^{+}_{abcd} + {}^{+}R^{-}_{abcd} = S_{a[d}g_{c]b} + S_{b[c}g_{d]a}\,, \tag{9.3}$$

where C_{abcd} is the Weyl tensor, and S_{ab} is the reduced Ricci tensor,

$$S_{ab} := R_{ab} - \frac{1}{4}Rg_{ab}\,. \tag{9.4}$$

Since F^{ab} is self-dual,

$$^{+}R_{abcd}F^{cd} = C_{abcd}F^{cd} - \frac{1}{6}RF_{ab}\,, \tag{9.5}$$

and

$$^{-}R_{abcd}F^{cd} = -F_{ap}S^{p}_{b} - S^{p}_{a}F_{pb}\,. \tag{9.6}$$

Taking the divergence of (8.19) on its last index, using (8.24) and (8.26), and remarking that

$$J^{d}{}_{;d} = 0 \tag{9.7}$$

identically, we obtain

$$F_{ab;cd}F^{cd} + \Phi F_{ab} = 0\,. \tag{9.8}$$

It now follows from (9.5) and the Ricci identity:

$$2F_{ab;[cd]} = F_{ap}R^{p}{}_{bcd} + F_{pb}R^{p}{}_{acd}\,, \tag{9.9}$$

that

$$F_{[a\,|\,p\,|}\,C^{p}{}_{b]qr}F^{rq} = \Phi F_{ab}\,. \tag{9.10}$$

By taking a basis for the self-dual bivectors, we find that the last equation implies that

$$C_{abcd}k^{b}F^{cd} = \Phi k_{a} \tag{9.11}$$

for any solution of (6.4): consequently,

$$R_{abcd}k^{b}F^{cd} = \Phi k_{a} - F_{ab}k_{c}S^{bc}\,, \tag{9.12}$$

from (9.5) and (9.6).

We now return to the deviation vector. Suppose, for the moment, that $t = 1$. From (8.29), writing

$$k_{a} = u_{,a}\,, \qquad l_{a} = tv_{,a}\,, \tag{9.13}$$

and using (6.17), we obtain

$$\Theta_{a}dx^{a} = \frac{1}{2}t(u^{;p}{}_{p}dv - v^{;p}{}_{p}du) \tag{9.14}$$

(Plebanski and Robinson, 1976): hence

$$\Theta_a = \frac{1}{2} t(v^{,p} u_{;pa} - u^{,p} v_{;pa}) , \tag{9.15}$$

because from (6.21) and (6.25),

$$(u_{,a} v^{,p} - v_{,a} u^{,p})_{;p} = 0 . \tag{9.16}$$

These equations also hold for arbitrary t, since they are invariant under the transformations (6.22) and (8.6). From (6.27) and (9.15),

$$u^{,p} u_{;pa} = 0 = v^{,p} v_{;pa} ,$$

$$t v^{,p} u_{;pa} = \Theta_a = -t u^{,p} v_{;pa} . \tag{9.17}$$

Putting $t = 1$ again, differentiating (9.15) covariantly, and writing

$$W_{ab} := u_{;a}{}^{p} v_{;pb} - v_{;a}{}^{p} u_{;pb} , \tag{9.18}$$

we get

$$\Theta_{a;b} = \frac{1}{2} (W_{ab} + v^{,p} u_{;pab} - u^{,p} v_{;pab}) ; \tag{9.19}$$

and hence, from the Ricci identity,

$$2\Theta_{[a;b]} = W_{ab} + \frac{1}{2} R_{abcd} F^{cd} . \tag{9.20}$$

Multiplying this equation by Θ^b, using (9.12), and considering that

$$W_{ab} \Theta^b = 0 , \tag{9.21}$$

from (9.14) and (9.17), we get

$$\Theta_{a;b} \Theta^b = \frac{1}{2} \Phi \Theta_a - \frac{1}{2} F_{ab} \Theta_c S^{bc} . \tag{9.22}$$

It follows, on account of (8.27) and (8.31), that

$$\Omega_a = \frac{1}{2} (\Phi \Theta_a + F_{ab} S^{bc} \Theta_c) . \tag{9.23}$$

This, as we shall see, is an interesting equation. We can derive it, without referring to the Weyl tensor, from a result of R. K. Sachs's (1962): for any l_a subject to (1.6),

$$\rho_{,q} l^q + \frac{1}{2} \rho^2 = \bar{\rho}_{,q} l^q + \frac{1}{2} \bar{\rho}^2 = S_{pq} l^p l^q . \tag{9.24}$$

Considering that du, dv, and $du + dv$ are shear-free on one side and geodesic, we have

$$u_{;p}{}^{pq}u_{,q} + \frac{1}{2}(u_{;p}{}^{p})^2 = S^{pq}u_{,p}u_{,q}\,, \tag{9.25}$$

$$v_{;p}{}^{pq}v_{,q} + \frac{1}{2}(v_{;p}{}^{p})^2 = S^{pq}v_{,p}v_{,q}\,, \tag{9.26}$$

and

$$u_{;p}{}^{pq}v_{,q} + v_{;p}{}^{pq}u_{,q} + u_{;p}{}^{p}v_{;q}{}^{q} = 2S^{pq}u_{,p}v_{,q}\,. \tag{9.27}$$

From (8.27) and (9.14), however, with $t = 1$,

$$u_{;p}{}^{pq}v_{,q} - v_{;p}{}^{pq}u_{,q} = 2\Phi\,. \tag{9.28}$$

Hence

$$u_{;p}{}^{pq}v_{,q} + \frac{1}{2}u_{;p}{}^{p}v_{;q}{}^{q} = S^{pq}u_{,p}v_{,q} + \Phi\,, \tag{9.29}$$

$$v_{;p}{}^{pq}u_{,q} + \frac{1}{2}u_{;p}{}^{p}v_{;q}{}^{q} = S^{pq}u_{,p}v_{,q} - \Phi\,. \tag{9.30}$$

Using (9.14) and (9.17), we obtain

$$\Theta_{a}{}^{;b}u_{,b} = \frac{1}{2}\Phi u_{,a} - \frac{1}{2}F_{ab}S^{bc}u_{,c} - u_{;b}{}^{b}\Theta_{a}\,, \tag{9.31}$$

from (9.25) and (9.30);

$$\Theta_{a;}{}^{b}v_{,b} = \frac{1}{2}\Phi v_{,a} - \frac{1}{2}F_{ab}S^{bc}v_{,c} - v_{;b}{}^{b}\Theta_{a} \tag{9.32}$$

from (9.26), (9.29); and hence (9.23).

In all this, of course, we are assuming that the distortion vanishes:

$$\Sigma_a = 0\,. \tag{9.33}$$

From (8.10), therefore,

$$2\Theta_{[a}F_{b]c} + F_{ab}\Theta_c = 0\,; \tag{9.34}$$

and from the skew product of (9.23) with Θ_b,

$$4\Theta_{[a}\Omega_{b]} = -F_{ab}S_{pq}\Theta^p\Theta^q\,. \tag{9.35}$$

The deviation vector Θ_a is always left-geodesic; it is right-geodesic if and only if the right expansion Ω_a is proportional to it. Thus from (9.35)

$$S_{pq}\Theta^p\Theta^q = 0 \tag{9.36}$$

is a necessary and sufficient condition for Θ_a to be geodesic.

10. A BIVECTOR BASIS

We can simplify some of the calculations by introducing a basis for the self-dual bivectors. For this purpose, we take a second left-handed null plane. A bivector is uniquely determined in this plane by the condition

$$\frac{1}{2} D_{ab}F^{ab} = 1 . \tag{10.1}$$

Writing

$$E_{ab} := 2D_{[a\,|\,p\,|}\, F^{p}{}_{b]} , \qquad \psi_c := D^{ab}F_{ab;c} , \tag{10.2}$$

we have

$$\begin{aligned} D_{ab}E^{b}{}_{c} &= D_{ac} , \qquad D_{ab}F^{b}{}_{c} = -\frac{1}{2}\, g_{ac} + \frac{1}{2}\, E_{ac} , \\ E_{ab}E^{b}{}_{c} &= g_{ac} , \qquad E_{ab}F^{b}{}_{c} = F_{ac} , \end{aligned} \tag{10.3}$$

and

$$F_{ab;c} = -E_{ab}\Theta_c + F_{ab}\psi_c \tag{10.4}$$

(Robinson and Schild, 1963). Transvecting (8.14) with D_{ab} and using (10.2), we obtain

$$\psi_c = \Psi_c - 2D_{ab}\Theta^b \tag{10.5}$$

in the absence of distortion: in the more general case, we shall take this as the definition of Ψ_c.

We select the plane of D^{ab}, quite arbitrarily, from a one-parameter family of left-handed null planes. Under a change of plane

$$D_{ab} \rightarrow D_{ab} + \chi E_{ab} + \chi^2 F_{ab} , \tag{10.6}$$

where χ is arbitrary, and

$$\Psi_a \rightarrow \Psi_a - 4\chi D_{ab}\, \Sigma^b - 2\chi^2 \Sigma_a . \tag{10.7}$$

Thus Ψ_a is invariant only when the distortion vanishes.

We use the basic bivectors, D^{ab}, E^{ab}, F^{ab}, to investigate the self-dual part of the Weyl tensor. This is most conveniently characterized by the quadratic form

$$\Gamma := \frac{1}{2}\, C_{pqrs}X^{pq}X^{rs} , \tag{10.8}$$

where X^{ab} is a null self-dual bivector. We may write it as

$$X^{ab} = \xi^2 D^{ab} + \xi\eta E^{ab} + \eta^2 F^{ab}, \tag{10.9}$$

thereby obtaining an expression for Γ of the form

$$\Gamma = C^{(1)}\xi^4 - 4C^{(2)}\xi^3\eta + 6C^{(3)}\xi^2\eta^2 - 4C^{(4)}\xi\eta^3 + C^{(5)}\eta^4. \tag{10.10}$$

Considering the symmetries of the Weyl tensor, we then have

$$C_{pqrs}F^{rs} = C^{(5)}D_{pq} + C^{(4)}E_{pq} + C^{(3)}F_{pq}, \tag{10.11}$$

$$-\frac{1}{2}C_{pqrs}E^{rs} = C^{(4)}D_{pq} + C^{(3)}E_{pq} + C^{(2)}F_{pq}. \tag{10.12}$$

Substituting into (9.10), we obtain

$$C^{(5)} = 0, \qquad C^{(4)} = \Phi, \tag{10.13}$$

and hence

$$C_{pqrs}F^{rs} = \Phi(g_{pq} + 2D_{pr}F^{r}{}_{q}) + C^{(3)}F_{pq}. \tag{10.14}$$

Equation (9.11) is an immediate consequence of this result.

From these equations and the algebraic properties of the basic bivectors, we obtain

$$F^{ap}F^{qr}C^{b}{}_{pqr} = \Phi F^{ab}, \tag{10.15}$$

$$(E^{ap}F^{qr} + F^{ap}E^{qr})C^{b}{}_{pqr} = -\Phi(E^{ab} + 2g^{ab}) - 3C^{(3)}F^{ab}. \tag{10.16}$$

For the sake of simplicity, we impose here the normalization (6.25). We then have

$$\Theta_a = E_{ab}\Theta^b = F_{ab}\psi^b \tag{10.17}$$

from (10.3) and (10.4). Hence, using (10.4) again, together with (10.11), (10.12), (10.13), we obtain

$$(F^{ap}F^{qr})_{;b}C^{b}{}_{pqr} = 5\Phi\Theta^a. \tag{10.18}$$

From (10.18) and the divergence of (10.15),

$$F^{ap}F^{qr}C^{b}{}_{pqr;b} = F^{ab}\Phi_{,b} - \Phi(J^a + 5\Theta^a). \tag{10.19}$$

This result was derived with $J^a = 0$, but we can easily verify that it is invariant under (3.1).

11. WEYL TENSOR AND FIELD EQUATIONS

At any point at which the self-dual part of the Weyl tensor is nonzero, it determines four left-handed null planes in tangent space. The corresponding bivectors are the roots of the equation $\Gamma = 0$: these are the principal null bivectors of the self-dual Weyl tensor. The anti-self-dual part of the Weyl tensor determines similarly four right-handed null planes. In the real case, planes of opposite chirality occur in complex conjugate pairs, and their real intersections are the four principal null directions of the Weyl tensor.

In the course of evaluating Ω^a, we have established that F^{ab} is a root of $\Gamma = 0$ and that it is a multiple root if and only if $\Phi = 0$. We shall now show that

$$\Phi = 0 \Leftrightarrow S^a = 0\,, \tag{11.1}$$

where

$$S^a := F^{ap}F^{qr}S_{pq;r}\,. \tag{11.2}$$

From the Bianchi identities,

$$C^b{}_{pqr;b} = S_{p[q;r]} + \frac{1}{12}\, g_{p[q}R_{,r]}\,, \tag{11.3}$$

and hence

$$S^a + \Phi(J^a + 5\Theta^a) = F^{ab}\Phi_{,b} \tag{11.4}$$

on account of (10.19). Here, again, it is convenient to put $J_a = 0$. We then get

$$F^{ab}\Psi_b = 3\Theta^a\,, \qquad \Theta^a\Psi_a = 0\,, \tag{11.5}$$

from (8.16);

$$\Psi_a S^a + 3\Phi_{,a}\Theta^a = 0\,, \tag{11.6}$$

by transvecting (11.4) with Ψ_a; and

$$\Phi^2 = \frac{1}{3}S^a\Psi_a - \frac{1}{5}S^a{}_{;a}\,, \tag{11.7}$$

from the divergence of (11.4), together with (8.27) and (11.6). From (11.4) and (11.7), (11.1) follows immediately. This result forms part of the generalized Goldberg-Sachs theorem (Robinson and Schild, 1963) mentioned above.

From (11.1) and the identities

$$F^{ap}S_{pq}F^{qb}\Psi_b = F^{ap}S_{pq}(3\Theta^q + J^q)\,, \tag{11.8}$$

$$(F^{ap}S_{pq}F^{qb})_{;b} = F^{ap}S_{pq}(\Theta^q + 2J^q) + S^a, \qquad (11.9)$$

there follows an important theorem of Plebanski and Hacyan (1975): if the *surface equations*

$$F^{ap}S_{pq}F^{qb} = 0 \qquad (11.10)$$

are satisfied, then

$$\Phi = 0. \qquad (11.11)$$

Hence, using (9.23), we obtain

$$\Omega_a = 0 \qquad (11.12)$$

as a consequence of the surface equations.

All this remains true in the special case defined by

$$\Theta_a = 0, \qquad (11.13)$$

but there we can derive more stringent restrictions on the curvature tensor. From (9.17), we see that the tensors $u_{;ab}$ and $v_{;ab}$ can be expanded in Cartesian products of $u_{,a}$ and $v_{,a}$: hence, from (9.18)

$$W_{ab} = 0, \qquad (11.14)$$

and from (9.20)

$$R_{abcd}F^{cd} = 0, \qquad (11.15)$$

as was shown by Plebanski and Robinson (1976). The self-dual part of this equation is

$$C_{abcd}F^{cd} = \frac{1}{6}RF_{ab}, \qquad (11.16)$$

which is equivalent to

$$C^{(5)} = C^{(4)} = 0, \qquad C^{(3)} = \frac{1}{6}R: \qquad (11.17)$$

thus F^{ab} is a principal null bivector of multiplicity three or more if and only if $R = 0$. The anti-self-dual part of (11.15) is equivalent to (11.10), its Maxwell product with F_{ab}.

In the more special case given by (11.13) and

$$\Psi_{[a,b]} = 0, \qquad (11.18)$$

we have

$$F_{p[a}R^p{}_{b]cd} = 0, \qquad (11.19)$$

from (8.17) and the Ricci identities. This decomposes into a self-dual part, with respect to the last pair of indices,

$$C^{(5)} = C^{(4)} = C^{(3)} = C^{(2)} = 0\,, \tag{11.20}$$

$$R = 0\,, \tag{11.21}$$

and an anti-self-dual part,

$$F^{ap} S_{pb} = 0\,. \tag{11.22}$$

12. A SPECIAL COORDINATE SYSTEM

So far, we have been working with completely general coordinates. We made extensive use, however, of the scalars u and v, and we now associate with them another pair of scalars, x and y, by requiring that

$$g^{ab} v_{,a} x_{,b} = 0\,, \qquad \mathcal{M} := g^{ab} u_{,a} x_{,b} \neq 0\,, \tag{12.1}$$

$$g^{ab} u_{,a} y_{,b} = 0\,, \qquad \mathcal{N} := g^{ab} v_{,a} y_{,b} \neq 0\,. \tag{12.2}$$

Writing

$$(x^1,\ x^2,\ x^3,\ x^4) = (u,\ x,\ y,\ v)\,, \tag{12.3}$$

we have

$$\det \mid g^{rs} x^a{}_{,r} x^b{}_{,s} \mid \ = (\mathcal{M}\mathcal{N})^2\,, \tag{12.4}$$

from which it follows that the x^a are functionally independent.

We take the x^a as coordinates. Then

$$g^{ab} u_{,b} = \mathcal{M} \delta_2^a\,, \qquad g^{ab} v_{,b} = \mathcal{N} \delta_4^a\,, \qquad g = (\mathcal{M}\mathcal{N})^{-2}\,, \tag{12.5}$$

and hence

$$u_{;p}{}^{p} = -\mathcal{M}(\ln \mathcal{N})_{,2}\,, \tag{12.6}$$

$$v_{;p}{}^{p} = -\mathcal{N}(\ln \mathcal{M})_{,4}\,. \tag{12.7}$$

All this arises from the existence of a bivector field F_{ab} that is simple, self-dual, and surface-forming. Suppose further that it satisfies the surface equations (see 11.10). From (11.12) and (8.32), it follows that $\Theta_a \propto V_{,a}$ for some scalar V. According to (9.14), V is a function of u and v only, and an appropriate transformation of the form (6.26) eliminates the dependence on u. Thus

$$v_{;p}{}^{p} = 0 \tag{12.8}$$

from (9.14), and

$$g^{ab}(u_{;p}{}^{p})_{,a}v_{,b} = 0\,, \tag{12.9}$$

$$g^{ab}(u_{;p}{}^{p})_{,a}u_{,b} + \frac{1}{2}(u_{;p}{}^{p}) = 0 \tag{12.10}$$

from (9.29), (9.25), (11.11), and the surface equations.

From (12.7) and (12.8), $\mathcal{M}_{,4} = 0$. Considering that (12.1) is invariant under the transformation $x \rightarrow \hat{x}(u, v, x)$, with $x_{,2} \neq 0$, we can write

$$\mathcal{M} = 1\,. \tag{12.11}$$

From (12.9) and (12.10), it follows that (12.1) and (12.11) are satisfied by Sachs's (1962) parameter

$$x = 2(u_{;p}{}^{p})^{-1}\,, \tag{12.12}$$

unless

$$u_{;p}{}^{p} = 0\,. \tag{12.13}$$

Thus we have

$$u_{;p}{}^{p} = 2(\ln \mathcal{X})_{,2}\,, \tag{12.14}$$

and hence

$$\Theta_a = (\ln \mathcal{X})_{,2}tv_{,a} \tag{12.15}$$

with

$$\mathcal{X} = x \text{ or } 1\,. \tag{12.16}$$

From (12.6), (12.11), and (12.14), $(\mathcal{N}\mathcal{X}^2)_{,4} = 0$; hence, by a transformation of the coordinate y, we obtain

$$\mathcal{N} = \mathcal{X}^{-2}\,. \tag{12.17}$$

Putting these results together, we have

$$g^{1b} = \delta_2^b\,, \qquad g^{a3} = \mathcal{X}^{-2}\delta_4^a\,. \tag{12.18}$$

That makes seven equations in all, leaving undetermined three components of the metric. The line element may be written as

$$ds^2 = d\hat{s}^4 + \mathcal{A}du^2 + 2\mathcal{B}dudv + \mathcal{C}dv^2\,, \tag{12.19}$$

where $\mathcal{A}$, $\mathcal{B}$, $\mathcal{C}$ are the undetermined functions, and

$$d\hat{s}^2 = 2dudx + 2\mathcal{X}^2dvdy\,. \tag{12.20}$$

For any metric of this form, writing

$$k_a := u_{,a} - \mathscr{X}^2 y \Theta_a , \qquad l_a := \mathscr{X} v_{,a} , \tag{12.21}$$

and putting $t = 1$ in (6.21) and (12.15), we have

$$k^a{}_{;q} F^{qr} = 0 , \qquad l^a{}_{;q} F^{qr} = 0 , \tag{12.22}$$

$$2 k^{[a} l^{b]} = \mathscr{X} F^{ab} \neq 0 . \tag{12.23}$$

The surfaces of constant u and v are thus spaces of distant parallelism: that is, in an affine sense, planes. We remark, incidentally, that

$$d\hat{s}^2 = 2kdx + 2ldz \tag{12.24}$$

where

$$k := k_a dx^a , \qquad l := l_a dx^a , \qquad z = \mathscr{X} y . \tag{12.25}$$

Suppose, conversely, that there exists a congruence of totally null planes, the space being otherwise unrestricted. Normalize the tangent bivector so that (6.25) holds. We have, by hypothesis, a scalar and a pair of vectors subject to (12.22) and (12.23). Taking the covariant divergence of (12.22), and using (6.25), we have

$$k^a{}_{;qr} F^{qr} = 0 , \qquad l^a{}_{;qr} F^{qr} = 0 , \tag{12.26}$$

whence

$$k^p R^a{}_{pqr} F^{qr} = 0 , \qquad l^p R^a{}_{pqr} F^{qr} = 0 , \tag{12.27}$$

from the Ricci identity, and

$$F^{ap} R^b{}_{pqr} F^{qr} = 0 , \tag{12.28}$$

from (12.23). Taking the trace-free symmetrical part of (12.28), we obtain the surface equations; and we can therefore specialize the coordinates so that (12.19), (12.20), and (12.16) hold.

Leaving aside the question of coordinates, we have here a theorem due to R. Penrose (personal communication, 1978) on a spacetime admitting a congruence of totally null surfaces: the surface equations are necessary and sufficient conditions for the surfaces to be flat.

The stronger equation (11.15) is necessary and sufficient for k and l to form part of a rigid tetrad covariantly constant on each of the null surfaces.

13. CONCLUSION

The background metric defined in (12.19) and (12.20) is flat, and the forms du and dv are null. Consequently, the full metric of (12.19) is a double Kerr-Schild metric. More specifically, it is the most general double Kerr-Schild metric with integrable two-spaces. It is this form that Plebanski and Schild (1976) picked out at an early stage of their research on account of the relatively simple form of its field equations. We can now describe it more directly, without reference to the unphysical background metric: a Plebanski-Schild space is one admitting a congruence of totally null planes.

Until about 20 years ago, if one wished to simplify the field equations in order to obtain exact solutions, one looked at spaces with some kind of symmetry. Then a succession of different possibilities was investigated with some measure of success, on the basis of ideas developed in the study of gravitational radiation.

Initially came the spacetimes admitting a congruence of null shear-free geodesics, soon identified as those for which the principal null directions of the Weyl tensor are not all distinct. Then came Kerr-Schild (1965) space, which turned out to be a special case of algebraically degenerate spacetime. More recently there has been systematic study of complex spacetimes with Weyl tensors that are self-dual or anti-self-dual (see in particular Newman, 1975; Penrose, 1976; and Plebanski, 1975). Finally, we have the double Kerr-Schild spaces or rather the special case we have described as Plebanski-Schild space. The field equations in all these classes of spacetime have proved to be more or less tractable. All these spacetimes have a common geometrical feature: they contain a congruence of totally null surfaces.

The question remains of what we are to do about the spaces that lack these surfaces, and of course, almost all spaces fall into this category. One possibility is to take very seriously the original idea of Kerr and Schild—the classification of metrics in terms of their relation to a flat background. For example, one might look at the reduced characteristic equation of the matrix

$$H^a{}_b := \eta^{ac} g_{cb} .$$

If this equation is of degree one, the space is conformally flat. If it is of degree two and if the roots are coincident, the space is conformal to a double Kerr-Schild space. If the coincident root takes a constant value, the space is double Kerr-Schild. This type of algebraic classification of metrics is a problem of no great difficulty.

How useful this classification program will be for finding solutions remains to be seen. Indeed, I should not like to leave the impression that we understand everything about the double Kerr-Schild metrics themselves. In the first place, there is the question of what happens when the tangent subspaces spanned by the two null vector fields are not integrable. It may be that this case is excluded by the field equations; otherwise, one may hope that the field equations reduce here to some moderately surveyable form. Should this case, which arises naturally in an algebraic classification scheme, not be tractable, then classification in terms of a flat background metric would be less interesting.

Even in the Plebanski-Schild case, in which these surfaces are integrable, interesting and important problems remain. In the form of line element given in (12.19) and (12.20), some of the field equations are built in. The others may be integrated explicitly to leave a single equation of second order and second degree (Plebanski and Robinson, 1977). All of this is highly satisfactory; unfortunately, the remaining equation is a singularly disagreeable one. We certainly cannot say that the problem is completely solved until we have understood this equation much better. After that, we shall be faced with the problem of identifying those solutions that have real cross-sections and putting the cross-sections into reasonable form.

Though problems remain, the Plebanski-Schild metrics represent an important milestone in the investigation of exact solutions to the field equations of general relativity.

The investigations described here have deepened our understanding of the geometry of spacetime and have produced a number of physically illuminating solutions. In all of this, Alfred Schild's contributions were of the greatest importance. He brought to this work not only vast technical skill but unusual mathematical and physical insight. His own achievements and his encouragement to others were responsible for a large part of the success attained.

REFERENCES

Cartan, E., "Sur les Espaces Conformes Généralisés et l'Univers Optique," *Comptes Rend. Acad. Sci. Paris* **174**, 857–860 (1922).

Goldberg, J., and Sachs, R., "A Theorem on Petrov Types," *Acta Phys. Polon.* **22** Suppl., 13–23 (1962).

Kerr, R. P., "Gravitational Field of a Spinning Mass as an Example of Algebraically Special Metrics," *Phys. Rev. Lett.* **11**, 237–238 (1963).

Kerr, R. P., and Schild, A., "A New Class of Vacuum Solutions of the Einstein Field Equations," in *Atti del Convegna sulla Relatività Generale: Problemi Dell'Energia e Onde Gravitazionali* (G. Barbera, Florence, Italy, 1965), pp. 222–233.

Lichnerowicz, A., "Radiations en Relativité Générale," in *Les Théories Relativistes de la Gravitation* (Cent. Nat. Recherche Sci., Paris, 1962), pp. 93–106.

Newman, E. T., "The Bondi-Metzner-Sachs Group: Its Complexification and Some Related Curious Consequences," in *General Relativity and Gravitation* (edited by G. Shaviv and J. Rosen, John Wiley and Sons, New York, 1975), pp. 137–142.

Penrose, R., "Nonlinear Gravitons and Curved Twistor Theory," *Gen. Rel. Grav.* **7**, 31–52 (1976).

Pirani, F. A. E., and Schild, A., "Geometrical and Physical Interpretation of the Weyl Conformal Curvature Tensor," *Bull. Pol. Acad. Sci.* **9**, 543–547 (1961).

———, "Conformal Geometry and the Interpretation of the Weyl Tensor," in *Perspectives in Geometry and Relativity: Essays in Honor of Vaclav Hlavaty* (edited by B. Hoffmann, Indiana Univ. Press, Bloomington, 1966), pp. 291–309.

Plebanski, J. F., "Some solutions of Complex Einstein Equations," *J. Math. Phys.* **16**, 2395–2402 (1975).

Plebanski, J. F., and Hacyan, S., "Null Geodesic Surfaces and Goldberg-Sachs Theorem in Complex Riemannian Spaces," *J. Math. Phys.* **16**, 2403–2407 (1975).

Plebanski, J. F., and Robinson, I., "Left-degenerate Vacuum Metrics," *Phys. Rev. Lett.* **37**, 493–495 (1976).

———, "The Complex Vacuum Metric with Minimally Degenerated Conformal Curvature," in *Asymptotic Structure of Space-time* (edited by F. P. Esposito and L. Witten, Plenum Press, New York, 1977), pp. 361–406.

Plebanski, J. F., and Schild, A., "Complex Relativity and Double KS Metrics," *Nuovo Cimento* **35B**, 35–53 (1976).

Robinson, I., Robinson, J. R., and Zund, J. D., "Degenerate Gravitational Fields with Twisting Rays," *J. Math. Mech.* **18**, 881–892 (1969).

Robinson, I., and Schild, A., "Generalization of a Theorem by Goldberg and Sachs," *J. Math. Phys.* **4**, 484–489 (1963).

Sachs, R. K., "Distance and the Asymptotic Behaviour of Waves in General Relativity," in *Recent Developments in General Relativity* (Pergamon Press, New York, 1962), pp. 395–407.

Schild, A., "Conservative Gravitational Theories of Whitehead's Type," in *Recent Developments in General Relativity* (Pergamon Press, New York, 1962), pp. 409–413.

Sommers, P., "Properties of Shear-free Congruences of Null Geodesics," *Proc. Roy. Soc.* **A349**, 309–318 (1976).

3. Linearization Stability

DIETER BRILL

1. INTRODUCTION

Linearization Stability Defined

To investigate the linearization stability of a nonlinear equation means, roughly speaking, to assess the reliability of the linearized approximation. Only in the last five years has this important problem in general relativity been attacked with appropriate mathematical methods. Today the essential features of the linearization stability of the Einstein equations are understood, but they are available only in technical research papers. The present exposition is addressed to those who prefer to sacrifice some rigor for a more intuitive and physical treatment.

Stability in general refers to persistence of some property under certain perturbations. In the usual stability analysis (of bridges or accelerator orbits, for example), one discusses a property (e.g., boundedness) of *solutions* of some equation. By contrast, linearization stability refers to a property of the equations themselves. By way of illustration, suppose we want to show that Newtonian gravitation is a limiting case of general relativity. A straightforward, but very difficult method would be first to find a sufficiently general set of exact solutions of Einstein's equations and then to show that an appropriate limit of these corresponds to Newtonian gravitational fields (Path 1 in Figure 3.1). Instead, the usual procedure is to prove that Einstein's equations become the Newtonian gravitational equations in a suitable limit (Path 2 in Figure 3.1). However, this second procedure does not show that there is also a correspondence between solutions. One needs to know in addition that solving and linearizing "commute." If they do, we say that the nonlinear equations are "linearization stable" (or "stable" for short). In particular, linearization stability implies that there are exact solutions, which the linearized solutions approximate. Proving stability is therefore one step in showing the existence of a variety of exact solutions.

Example of an Unstable Equation

Instability of an equation does not necessarily make it atypical or unphysical. An example of a highly unstable but physically quite meaningful equation is the minimal surface equation in three-dimensional Euclidean space. It describes, for example, the possible configurations of (static) soap films. For the surface represented by $z = f(x,y)$ this equation takes the form

$$(1 + |\,\mathrm{grad}\, f\,|^2)\,\Delta f - \mathrm{Hess}\, f(\mathrm{grad}\, f, \mathrm{grad}\, f) = 0\,, \tag{1}$$

where Δ is the two-dimensional Laplacian and $(\mathrm{Hess}\, f)_{AB} = \partial^2 f/\partial x^A \partial x^B$, where $A,B = 1,2$. Suppose we are looking for solutions f that are defined for all x,y. Let $f_\lambda(x,y)$ be a family of solutions with $f_0 \equiv 0$, and denote the linear approximation by $g(x,y) = \partial f_\lambda(x,y)/\partial\lambda\,|_{\lambda=0}$. Equation (1) is then easily linearized (by differentiating with respect to λ and then setting $\lambda = 0$) to give Laplace's equation for g,

$$\Delta g = 0\,. \tag{2}$$

This equation in turn is easily solved:

$$g = \text{harmonic function} \tag{3}$$

(e.g., $ax + by$, $x^2 - y^2$, etc.).

To solve (1) first is more difficult, but the result is that the general solution has the *linear* form (Bernstein's theorem),[1]

$$f(x,y) = ax + by\,. \tag{4}$$

Hence the linear approximation to a general solution (obtained, e.g., from the family $f_\lambda(x,y) = \alpha(\lambda)x + \beta(\lambda)y$ is also a linear function (with $a = \alpha'(0)$, $b = \beta'(0)$)

$$g(x,y) = ax + by\,. \tag{5}$$

Thus all but the linear solutions of the linearized equations (2) fail to be approximations to solutions of the exact equation (1). This result shows that the minimal surface equation (1) is highly unstable at the solution $f_0(x,y) \equiv 0$. Moreover, we can choose the plane described by any exact solution (4) as the xy plane of a new coordinate system and repeat the argument. *Hence, the minimal surface equation (1) is linearization unstable at all of its solutions.*

Simplification by 3 + 1 Decomposition

Two features simplify the linearization stability analysis of Einstein's equations. The first is that stability is independent of the choice of metric variables. If we make a regular change of the variables used to describe the metric, then a linear approximation in terms of the old variables becomes, after change of metric variables, a linear approximation in terms of the new variables. Therefore, it does not matter how we describe the metric in order to

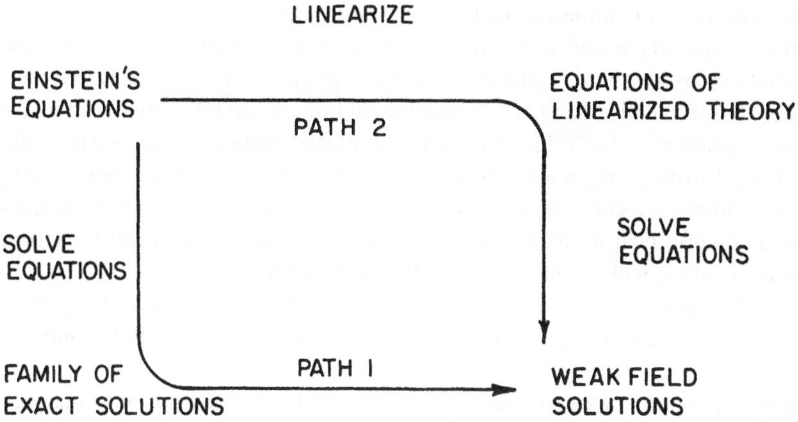

Figure 3.1. Linearization stability illustrated by "weak field approximation" to Einstein's theory. To be sure that a weak field solution has something to do with exact solutions of Einstein's equations, the weak field solution should be obtained as a limit of exact solutions (Path 1). Only if linearizing and solving are interchangeable can weak field solutions correctly be obtained from the linearized equations (Path 2). Linearization stability assures this interchangeability.

discuss the reliability of the linear approximation. (In the language of the following section on geometrical formulation, we are merely describing the smoothness of the same subset of solutions in terms of different variables, related by smooth transformations.) This allows us, for example, to replace the ${}^4g_{\mu\nu}$ by the "3 + 1 decomposition" ${}^3g_{ij}$ (metric induced on x^0 = const hypersurface) and N, N_i (lapse and shift functions) with respect to a foliation by spacelike hypersurfaces (Misner, Thorne, and Wheeler, 1973).

The second simplifying feature is that we need to consider only the stability of the initial value equations. These equations tell us all we need, because a solution of the initial value equations determines a unique solution (up to diffeomorphism) of the full Einstein equations (Hawking and Ellis, 1973). Thus there is a unique correspondence between all the relevant three-dimensional and four-dimensional quantities (such as spaces of exact solutions, their tangent spaces, etc.). We shall therefore work primarily with the initial value equations, not only because they are fewer in number but, more important, because they can be cast into elliptic form. (It turns out that standard methods, such as Banach space analysis, apply most naturally to elliptic systems.) Interestingly enough, many of the final results are most simply stated in their four-dimensional, spacetime version, but present methods of proof have to resort to the 3 + 1 decomposition.

Geometrical Formulation of Linearization Stability

It is frequently useful to think of metrics on a given manifold M, whether solutions of Einstein's equations or not, as points in a large (infinite dimensional) space called Lor (M). (The notation Lor (M) signifies that each metric has signature $(-1,1,1,1)$, called the Lorentzian signature.) The metrics that satisfy Einstein's equations are a subset of Lor (M). A one-parameter family of solutions is a curve that lies in the solution subset. A linear approximation to a solution near a given solution then corresponds to the tangent vector to such a curve, which passes through the given solution.

The problem of linearization stability can be formulated in this general context. By linearizing Einstein's equations about any given solution and then solving, we obtain a linear space at any point in the solution subset. If this linear space coincides with the tangent space to the solution subset—that is, if each solution of the linearized theory is tangent to a family of exact solutions—then the Einstein equations, which define the solution subset, are stable at this point in Lor (M). Stability means that a tangent space exists and that the solution subset is a smooth manifold at this point (Figure 3.2). Because all of classical general relativity is represented by the solution subset, it is clearly desirable to explore as many of its properties as one can. One can hardly think of a more fundamental property than the existence of a tangent space as determined by the stability of Einstein's equations.

This general view of Einstein's equations allows us to illustrate the idea of stability by means of a finite-dimensional example. Let the space be three-dimensional and the equation be[2] (Figure 3.3)

$$F(X,Y,Z) = X^2 + Y^2 - Z^2 = 0\,. \tag{6}$$

The space of solutions is a double cone. The linearized equation is, near any exact solution X_p, Y_p, Z_p (we let $'$ denote $d/d\lambda$),

$$2X_pX' + 2Y_pY' - 2Z_pZ' = 0\,. \tag{7}$$

If $X_p, Y_p, Z_p \neq 0$, the space of X', Y', Z' satisfying (7) is two-dimensional and describes the tangent space to the cone. However, at $X_p = Y_p = Z_p = 0$, equation (7) is satisfied by the unrestricted, three-dimensional space of X', Y', Z'. This result is not surprising, for (a) all tangents to the cone at the origin must be in the space of solutions of the linearized equations; (b) this latter space must be a linear space; (c) the tangents to the cone at the origin do not form a linear space, but their linear combinations span a three-dimensional space. Many directions in this linear space are not tangent to the cone. Thus the equation $F(X,Y,Z) = 0$ is unstable at the origin and stable everywhere else.

The above example suggests the possibility that the true tangent directions in the unstable case might be found by approximating the equation to a higher order. This procedure is indeed valuable, but its success depends on the

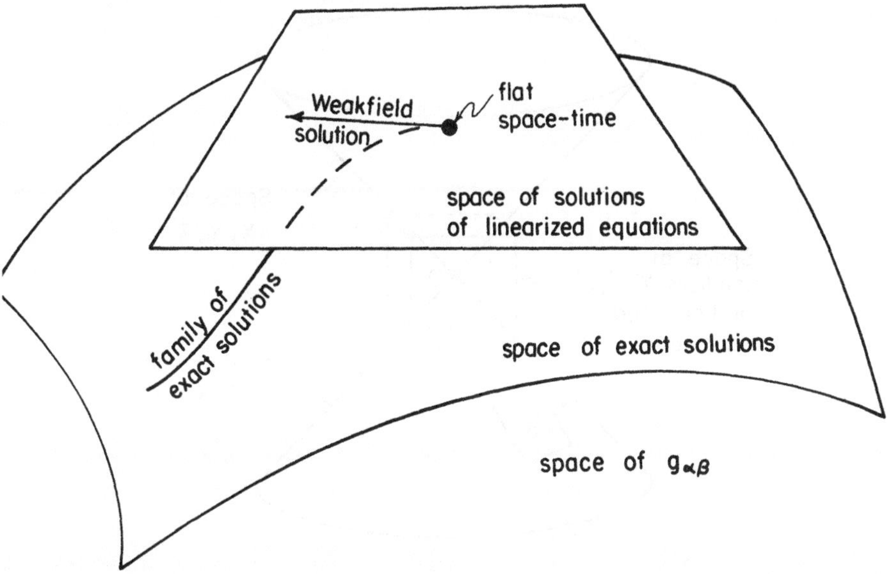

Figure 3.2. Relation between exact and linearized equations. The figure shows a space whose points represent metrics on a given manifold. Those metrics that satisfy Einstein's equations lie in the "space of exact solutions." A one-parameter family of solutions near flat spacetime has a weak field solution as its tangent. If Einstein's equations are linearization stable at flat space, the space of exact solutions is a smooth manifold at that point, and its tangent space is the same as the space of solutions of the linearized equations. The stable case is shown here. An unstable situation is shown in Figure 3.3.

particular equation. Clearly, second order will work in the above example, as indicated in the legend of Figure 3.3. But the lowest order that gives nontrivial equations is not generally sufficient, as shown by the equation $x^4 + y^6 = 0$. Finally, we mention that although stability of the equation always implies that the solution space is regular, the converse is not true; a counterexample is $(x - y)^2 = 0$, which is unstable at *every* exact solution but which nonetheless describes a smooth manifold.

We can now give a plausibility argument to characterize the unstable situation. At an instability we expect second or higher order conditions on the solutions of the linearized initial value equations (cf. Figure 3.3). Since instability is a property of the spacetime, it should occur in the initial value problem on all Cauchy surfaces. Second order conditions on each surface would be too great a restriction, unless the conditions are automatically satisfied on all Cauchy surfaces if they are satisfied on one surface. Hence we ex-

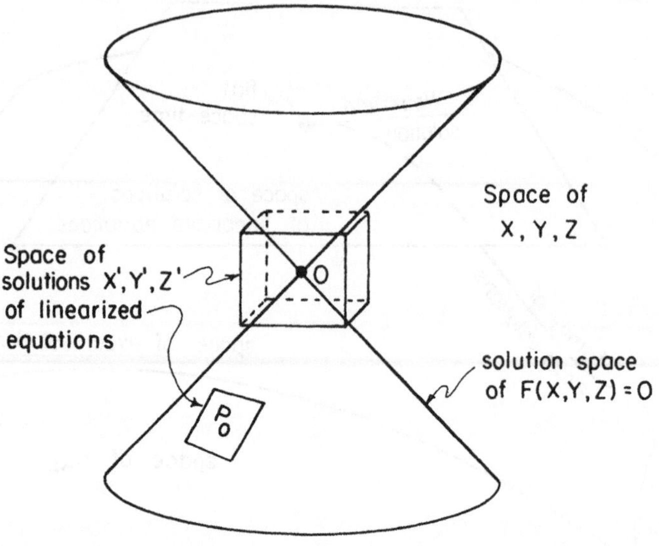

Figure 3.3. Instability of $F(X,Y,Z) = X^2 + Y^2 - Z^2 = 0$. At any point P different from the origin O, the linearized equation $dF/d\lambda|_{\lambda=0} = 2X_pX' + 2Y_pY' - 2Z_pZ' = 0$ correctly describes the tangent plane. At the origin the linearized equation is satisfied by a three-dimensional space of (X', Y', Z') (symbolized by a cube). Many directions in this space are not tangent to the solution space, thus the equation $F(X,Y,Z) = 0$ is unstable at the origin: the linearized solutions there (and only there) have to be subjected to the additional, second order condition $F''(O) = (X')^2 + (Y')^2 - (Z')^2 = 0$.

pect the second or higher order conditions to be conserved quantities, which we in turn expect to be associated with Killing vectors. We shall see below that in closed spaces instability indeed occurs precisely where the "unperturbed" spacetime geometry admits Killing vectors. By contrast, in asymptotically flat spacetimes, the conservation laws also allow surface integrals at infinity, and we shall see that asymptotically flat spacetimes are generally linearization stable.

2. THE IMPLICIT FUNCTION THEOREM

First Version of the Theorem

The initial value equations, or constraints, are the four Einstein equations, $G_\mu{}^0 = 0$. We will also write them as $H = 0$ ("Hamiltonian constraint" $G_0{}^0 = 0$) and $H^i = 0$ ("momentum constraints" $G_i{}^0 = 0$). They constrain the Cauchy data, which we take to be the three-dimensional metric g_{ij} and the "conjugate momenta," $\pi_{ij} = (K_{ij} - g_{ij}\, K_l^{\,l})$, with K_{ij} the second fundamental form of the Cauchy surface. Let $\mathcal{M} = \text{Riem}(M)$ be the space whose points are all possible positive-definite (Riemannian signature) metrics g_{ij} on a fixed three-dimensional manifold M. Its cotangent bundle $T^*\mathcal{M}$ is then the space of all pairs (g_{ij}, π_{ij}), since π_{ij} can be considered a directional (time) derivative of g_{ij} and hence a cotangent vector. The constraints define a subset C of $T^*\mathcal{M}$, called the constraint subset, on which these constraints are satisfied. (This is quite analogous to the solution subset in the spacetime context discussed above.) The definition of C is an implicit definition, and to make it explicit we would have to solve the constraints. The question of whether the constraints can be solved is therefore analogous in the finite dimensional case to the problem of solving explicitly for an implicitly defined function. Here the implicit function theorem is the appropriate tool, because it relates the existence of a solution to the behavior of the linearized approximation (i.e., the derivative) of the implicit function.

There are well-known prescriptions (Ō Murchadha and York, 1974*a*) that tell us which of the 12 functions g_{ij}, π_{ij} to choose arbitrarily and which of them are the "dependent variables" that are determined by the constraints. Let X denote the eight independent functions, and Y the dependent functions. (Usually one chooses for X the five conformal metric components, the two transverse traceless components of π_{ij}, and the trace $\pi = \pi_i^{\,i}$; for Y one chooses the conformal factor ψ and the three longitudinal parts $\pi^{\,L}_{ij}$ of π_{ij}.) The constraints then have the form of the implicit equation $F(X,Y) = 0$ and are to be solved for $Y = Y(X)$.

Consider first the analogous finite dimensional case, where X is an m-component vector, and Y and F are n-component vectors. To find zeros of the smooth function F, we can use a version of the iteration known as Newton's method. Let X_0, Y_0 be a solution, so that $F(X_0,Y_0) = 0$, and choose some fixed X near X_0. Define a sequence $\{Y_n\}$ by

$$\begin{aligned} Y_{n+1} &= Y_n - (D_Y F_0)^{-1} F(X,Y_n) \\ &= Y_0 - (D_Y F_0)^{-1} [F(X,Y_n) - D_Y F_0 (Y_n - Y_0)] \,. \end{aligned} \tag{8}$$

Here $D_Y F_0$ denotes the $n \times n$ Jacobian matrix $(\partial F_a(X,Y)/\partial Y_b)$ evaluated at X_0, Y_0. Figure 3.4 shows how the sequence of Y_n converges to a solution of $F(X,Y) = 0$ for a simple example.

If $D_Y F_0$ is nonsingular and X is sufficiently close to X_0, this sequence will always converge. Namely, we note that the bracket in equation (8) is quadratic (or higher) in Y about some point near Y_0. Hence the sequence $\{Y_n\}$ is a Cauchy sequence, which converges.

In the case of interest for general relativity, the vectors X and Y are functions, and the vector F is the left side of the constraint equations. In order to

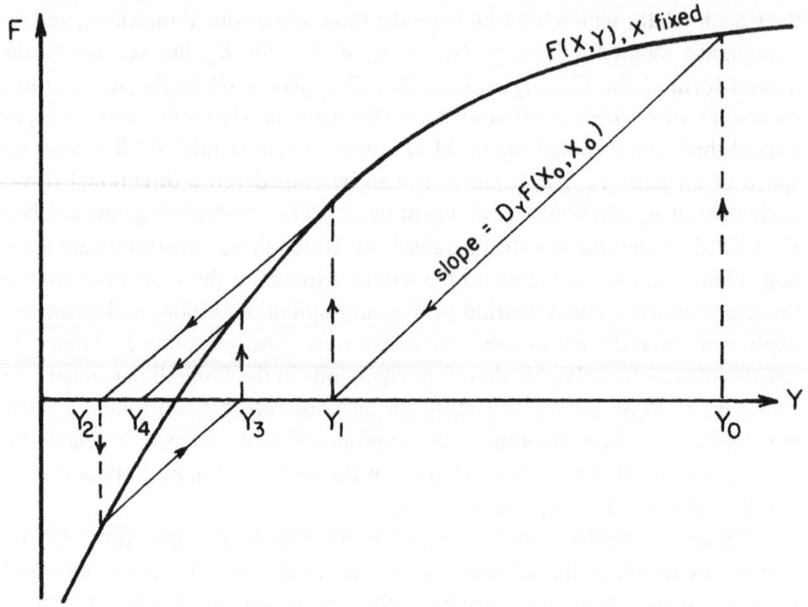

Figure 3.4. Modified Newton's method to find zero of $F(X,Y)$ for fixed X. To compute Y_{n+1} from Y_n by straight line extrapolation, we here do not use the "correct" slope at Y_n but the approximately correct and *constant* slope $D_Y F(X_0, Y_0)$ at a nearby solution X_0, Y_0. The distance between successive points, $|Y_{n+1} - Y_n|$, contracts by at least some factor $k < 1$ ($k \sim 0.8$ in the figure). Hence the sequence $\{Y_n\}$ is a convergent, Cauchy sequence.

apply the procedure of equation (8) we must (*a*) define when two vectors (such as X and X_0 or Y_n and Y_{n+1}) are close to each other; (*b*) give a definition of $D_Y F$; (*c*) make sure that equation (8) defines a Cauchy sequence; and (*d*) know that Cauchy sequences converge. Therefore, it is appropriate to require that X, Y, and F be points in complete, normed, linear function spaces (i.e., in Banach spaces) (Choquet-Bruhat, DeWitt-Morette, and Dillard-Bleick, 1977). The derivative $D_Y F$ can be defined in such spaces—it is just the linear operator one obtains when linearizing the expression F. The implicit function theorem then holds in Banach spaces (Choquet-Bruhat, DeWitt-Morette, and Dillard-Bleick, 1977). If F is a smooth map of elements X, Y in two Banach spaces into a third Banach space and if $D_Y F$ is an isomorphism at a solution X_0, Y_0 of $F(X_0, Y_0) = 0$, then for $|X - X_0|$ small there is a solution $Y = Y(X)$ of $F(X,Y) = 0$.

Showing linearization stability by this method therefore involves two steps: (*a*) find the appropriate Banach spaces and (*b*) show that $D_Y F$ is nonsingular. These steps are interdependent, and both have physical content. For example, requiring a finite Banach norm may impose asymptotic conditions such as asymptotic flatness; and whether the operator in the linearized equations is nonsingular may depend on the asymptotic conditions. A simple example will illustrate the problems involved.

A "One-dimensional" Example

We apply the procedure of equation (8) to the following situation: The "unperturbed" solution X_0, Y_0 is flat. The "first order perturbation" of the variable X is given by the Cauchy data of a weak gravitational plane wave,

$$h_{\mu\nu} = \mathrm{diag}(2\alpha(z - t),\ -2\alpha(z - t),\ 0,\ 0)\,, \tag{9}$$

where the function of one variable α specifies the shape of the weak wave. Since these $h_{\mu\nu}$ are transverse, traceless, and a solution of the (flat spacetime) wave equation, they represent a solution of the linearized equations (Misner, Thorne, and Wheeler, 1973). The corresponding initial values that agree with the above to lowest order in α are, with $\alpha'(z) = d\alpha(z)/dz$,

$$g_{ij} = \mathrm{diag}(e^{2\alpha(z)},\ e^{-2\alpha(z)},\ 1);\ \pi_{ij} = \mathrm{diag}(-\alpha'(z),\alpha'(z),0) \tag{10}$$

One can easily check that these initial values satisfy the linearized constraints. Also, inasmuch as det $(g_{ij}) = 1$ and the π_{ij} are transverse and traceless, these data have the form appropriate for the independent variables X, and the dependent variables Y have their flat-space values, $\psi_0 = 1$, $\pi^L_{ij(0)} = 0$.

The question of linearization stability is, Can we solve the exact constraints for ψ and π^L_{ij}? If we can do so, we can compute an exact solution of the constraints that corresponds to the given linearized solution, by the prescription (Ō Murchadha and York, 1974*a*),

$$g_{ij\,solution} = \psi^4 g_{ij};\ \pi_{ij\,solution} = \psi^{-2}\overline{\pi}_{ij} = \psi^{-2}(\pi_{ij} + \pi^L_{ij}).$$

Consider first the Hamiltonian constraint, which is to be solved for the conformal factor ψ. It has the form (for traceless π_{ij})

$$F(X,Y) = \psi^5 H(g,\pi;\psi) = 8\nabla_g^2\psi - R(g)\psi + \overline{\pi}_{ij}\overline{\pi}^{ij}\psi^{-7} = 0\,. \tag{11}$$

Here ∇_g^2 is the Laplacian in the metric g_{ij}, and $R(g)$ is the scalar curvature of this metric. Since the operator that acts on ψ in equation (11) is already linear when the X variables are evaluated at X_0 (i.e., flat g_{ij} and vanishing π_{ij}), we can immediately identify the derivative

$$D_Y F(X_0,Y_0) = D_\psi H(g_\flat,0;\psi)\,|_{\psi=1} = 8\nabla^2_\flat\,. \tag{12}$$

Here a subscript $\flat$ denotes flat-space quantities.

The first iteration of equation (8) starts with $\psi_0 = 1$. Because $R(g) = -2(\alpha')^2$ and $\overline{\pi}_{ij}\overline{\pi}^{ij} = 2(\alpha')^2$, we have

$$\psi_1 = 1 - (2\nabla^2)^{-1} [(\alpha')^2], \tag{13}$$

and similarly we would have for general n,

$$\psi_{n+1} = 1-(8\nabla^2)^{-1} [2(\alpha')^2\psi_n + \overline{\pi}_{ij}\, \overline{\pi}^{ij}\psi_n{}^{-7}] \tag{14}$$

We can now see that the iteration already fails at $n = 0$: Because $(\alpha')^2$ is a positive function of z only, $\int(\alpha')^2 dxdydz$ is not finite. But to satisfy equation (13) we need $(\alpha')^2 = -2\nabla^2\psi$, and $-\int 2\nabla^2\psi dxdydz = -2\int\nabla\psi\cdot n dS$ is finite for any physical ψ. Mathematically, the implicit function theorem fails here because $D_Y F = 8\nabla^2$ is not an isomorphism between the space of physically acceptable ψ's and the space of images of F, of which $(\alpha')^2$ is a member. Physically, we can interpret the difficulty as follows. Think of the term $F(X,Y_n) = H(g,\pi;\psi_n)$ in equation (8) as that part of the effective gravitational field's energy that has not yet been taken into account as a source of curvature. For the independent variables given by the plane wave of equation (11), the total (integrated) effective energy is infinite, and no finite conformal factor ψ can describe the response of the geometry to this source of curvature.

Thus one of the requirements on the independent variables X in order to generate a solution ψ is that at each stage of the iteration the total effective energy is finite,[3] and there may be additional conditions to make the iteration possible, such as finiteness of higher moments of the effective energy and momentum. A systematic way to state all these requirements is by the conditions of the implicit function theorem.

Thus, by choice of appropriate Banach spaces, we can make $D_Y F$ an isomorphism between the space of Y and the space of images $F(X,Y)$; and this choice has a direct physical meaning, at least in terms of this notion of effective gravitational energy: solutions of the linearized equations do not correspond to exact solutions if the effective energy is too large for the geometry to respond, by a small modification in higher orders, to this energy as a source. However, linearized solutions with too large an effective energy—as those of equation (10)—are *not* considered points of instability. They are simply not *small* perturbations. This is so because the extra conditions on the linearized solutions can be expressed as an inequality of the type $|X - X_0| < \varepsilon$, with $|\ |$ denoting the norm in a suitable Banach space.

Similar conditions arise from the momentum constraints. The data of equation (10) satisfy these constraints to lowest order (i.e., in flat space), but in higher order have to be modified (Ō Murchadha and York, 1974*a*) by a longitudinal part. This modification is determined by a vector λ (which now plays the role of the dependent variable Y) according to

$$\overline{\pi}_{ij} = \pi_{ij} + \lambda_{i\,|\,j} + \lambda_{j\,|\,i} - \frac{2}{3}\, g_{ij}\lambda^k{}_{|\,k}, \tag{15}$$

where | denotes covariant derivative in the metric g_{ij}. The momentum constraints are

$$H^i(g,\pi;\lambda) = \overline{\pi}^{ij}{}_{|j} = 0\,. \tag{16}$$

The linearization, expressed here as acting on a variation $\delta\lambda$, is

$$D_\lambda H^i(g_\flat,0;0)[\delta\lambda] = (\delta\lambda^{i|j}+\delta\lambda^{j|i} - \frac{2}{3}\, g_\flat^{ij}\delta\lambda^k{}_{|k})_{|j}\,. \tag{17}$$

The first iteration of equation (8) would give λ_1 from $\lambda_0 = 0$ by

$$\lambda_1 = -[D_\lambda H^i(g_\flat,0;0)]^{-1}\,[H^i(g,\pi;0)]\,, \tag{18}$$

with $H^i(g,\pi;0) = \pi^{ij}{}_{|j} = -2(\alpha')^2\delta^i_3$. Again we have $\int H^3(g,\pi;0)\, dxdydz = -2\int(\alpha')^2 dxdydz$ not finite because α' depends only on z and $(D_\lambda H^i)^{-1}$ is not defined on such functions. Physically, the difficulty can be characterized as a divergent total gravitational momentum that is unaccounted for in the linearized solution. The cure is again not to declare instability, but to choose the linearized g's and π's of sufficiently small norm in a suitable Banach space.

We have so far not specified what the "suitable" Banach space is, and it is perhaps not very surprising that several choices of the norm in the space of g's and π's are possible, with corresponding choices of norm in the space of images under the constraint expressions. Since no one choice has to date been recognized as clearly preferable, we mention here only that for stability a typical norm imposes smallness requirements on $g_{ij} - g_{ij\flat}$ and π_{ij} as well as on several derivatives and on their asymptotic fall-off. For example, π_{ij} should fall off as r^{-2}, and $R(g_{ij})$ as r^{-4}. For details we refer to the technical literature (Choquet-Bruhat and Deser, 1973; Ō Murchadha and York, 1974*b*; Choquet-Bruhat, Fischer, and Marsden, 1979; Cantor, 1977).

Example of Instability in Closed Flat Space

A simple modification of the above example can remove the divergence of the total effective gravitational energy and momentum. Choose a periodic function $\alpha(z)$, and identify points with this period in z-coordinate. Also make similar identifications in the x- and y-coordinates. The resulting topology is that of a closed three-dimensional torus. We can now follow the computations as above. Again the iteration fails at the first step, but this time because $\int(\alpha')^2 d^3x > 0$. Hence it cannot equal $-\int\nabla^2_\flat\psi d^3x$, which vanishes in a closed space. The momentum constraint similarly implies the vanishing of an integral. Here the failure to find an exact solution corresponding to our periodically identified linearized solution (10) *does* show that there is instability, because the second order condition—$\int(\alpha')^2 d^3x = 0$—is an *equality*, rather than an inequality as in the asymptotically flat case.

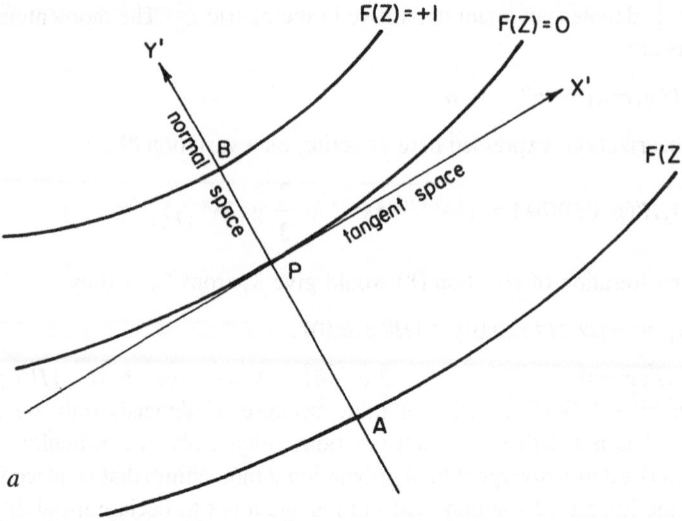

Figure 3.5. Second version of implicit function theorem illustrated. (*a*) Appropriate coordinates in which to solve $F(Z) = 0$ are tangent vectors X' (for which $DF[X'] = \text{grad } F \cdot X' = 0$) and normal vectors Y'. First version of implicit function theorem gives us criterion for smooth C at P that $(D_Y F)_P \neq 0$; hence DF must map vectors at P (tangent or normal) *onto* ("surjectively") a neighborhood of the origin of image space under F (i.e., the real numbers in case of the Figure). For example, vectors between A and B cover the interval $[-1,1]$ of the image space.

The flat three-dimensional torus is not a totally unstable (i.e., isolated) solution, because there are some linearized perturbations that do correspond to exact solutions. These perturbations satisfy the second order equalities and in addition, of course, some inequality (smallness) conditions of their Banach norm. (See "Example of Second Order Conditions: Perturbations of the Flat Three-dimensional Torus" below.)

Second Version of the Theorem

The first version of the implicit function theorem can be used to establish stability (if $D_Y F(X_0, Y_0)$ is nonsingular), but failure of this criterion does not, conversely, necessarily mean that there is instability. A simple example for functions is $F(X,Y) = X - Y^2 = 0$ near $X_0 = Y_0 = 0$. Here $D_Y F(X_0, Y_0) = 0$—and indeed one cannot solve uniquely for $Y = Y(X)$—but the equation defines a smooth manifold. An example in general relativity is the (vacuum) constraints near time-symmetric initial values on a closed space (e.g., the time-symmetric Taub universe). Here $\pi_{ij(0)}$ and $R(g_{ij(0)})$ both vanish, hence $(D_\psi H)_0 = 8\nabla^2_{g(0)}$, which is singular because it annihilates constant func-

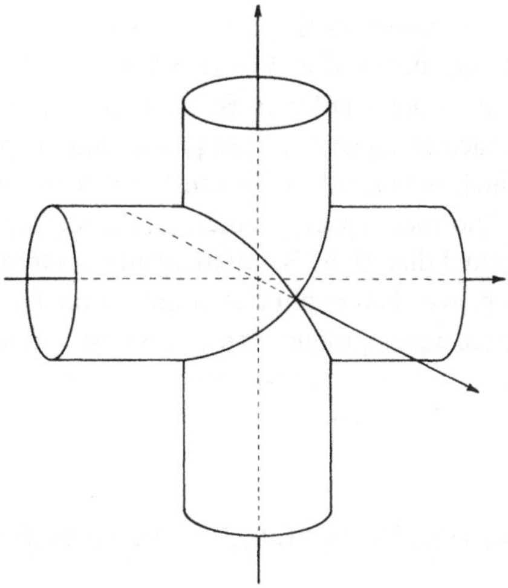

b

(*b*) The curve of the intersection of two cylinders illustrates this criterion. The curve is described implicitly by the simultaneous equations

$$U(x,y,z) = x^2 + y^2 - a^2 = 0\ ,\ V(x,y,z) = x^2 + z^2 - a^2 = 0.$$

Thus the map F maps $R^3 \rightarrow R^2$. Its Jacobian is $\begin{pmatrix} 2x & 2y & 0 \\ 2x & 0 & 2z \end{pmatrix}$.

Although this annihilates some vectors in R^3 (namely, tangents to the solution curve), the space of images at most points on C is R^2, the same as that of F. However, at $x = \pm a$, $y = z = 0$, the Jacobian is $\pm 2a\left(\begin{smallmatrix} 1 & 0 & 0 \\ 1 & 0 & 0 \end{smallmatrix}\right)$. Here all images are proportional to $\left(\begin{smallmatrix} 1 \\ 1 \end{smallmatrix}\right)$, so the image space is R^1, the map is not subjective, and there is instability at these points.

tions. However, the Hamiltonian constraint in some time-symmetric closed universes (e.g., the Taub universe) is stable.[4]

The reason for failure of the criterion in both cases is a bad choice of variables. In the case of $F(X,Y) = X - Y^2 = 0$, it is clearly more appropriate to choose Y as the independent variable and solve for $X(Y) = Y^2$. Similarly, in the general case it is appropriate to choose new variables so that the independent variables are tangent to the constraint manifold C, and the dependent variables are normal to it, at the unperturbed point (in some metric). The object is to use the first version of the implicit function theorem in the new variables and express the result in a way that is independent of the choice of variables. To do this, consider $F(X,Y) = F(Z)$ as a map of the total space of all X,Y, and let DF be its $n \times (n + m)$ Jacobian matrix $\partial F_a/\partial Z_B$ (or, in the case of function spaces, the linearized operator acting on all the variables). The result is that at Z_0, C is a smooth manifold and $F(Z) = 0$ is linearization stable, if $DF(Z_0)$ is a surjection (i.e., a map that "fills" the entire tangent space of 0 in the image space). (See Figure 3.5.)

Again the same theorem holds in Banach spaces, provided the tangent space to the total space at Z_0 splits, as required, into a space annihilated by DF (the kernel, which is tangent to the constraint manifold) and a topologically closed space ("normal" space). If there is a Hilbert space norm, the split into tangent and normal directions is automatically assured by the inner product. In the following, we shall confine attention to metrics on closed spaces, where a Hilbert space inner product can always be defined. Thus, we only need to check whether DF is surjective, without having to split variables into dependent and independent ones.

3. APPLICATIONS OF THE IMPLICIT FUNCTION THEOREM

The theorems of the preceding section have been used to investigate stability of a variety of nonlinear differential equations. The applications in general relativity of Choquet-Bruhat and Deser (1973), Ō Murchadha and York (1974*b*), Choquet-Bruhat, Fischer, and Marsden (1979), Cantor (1977), and others showing the stability of asymptotically flat space have already been discussed in outline. In the present section we follow the work of Fischer and Marsden (1975) and Moncrief (1975, 1976), to investigate the stability of closed spaces in general relativity. For simplicity we confine attention to the sourceless, vacuum case. The procedure is to use the second version of the implicit function theorem to show stability of a large class of solutions. For the exceptional cases (such as the flat three-dimensional torus already discussed), when the theorem fails, we show instability by giving nontrivial second order conditions. Thus it is possible to classify completely solutions of Einstein's equations according to their linearization stability.

Evaluation of the "Linearized Constraints," *DF*

A discussion of the "Jacobian" DH, DH_i of the constraints was first given in generality by Moncrief (1975, 1976) and considerably simplified by the work of Arms (1977) and Choquet-Bruhat (1977). We first introduce notation to allow us to distinguish between the Banach space $T^*\mathcal{M}$ of g_{ij}, π_{ij} ("phase space") of which the exact solutions are a subset C, and its tangent space at any point, $T(T^*\mathcal{M})$, which contains the solutions of the linearized equations. Let $g_{ij}(\lambda)$, $\pi_{ij}(\lambda)$ be a curve on $T^*\mathcal{M}$ and let $\delta \equiv d/d\lambda$. Elements of $T(T^*\mathcal{M})$ are $(\delta g_{ij}, \delta\pi_{ij})$, which we also denote by $\delta g_{ij} = h_{ij}$, $\delta\pi_{ij} = \omega_{ij}$. The image space $\mathcal{N}$ of the constraints consists of scalars (H) and vectors (H_i).[5] The linearized constraints map to the tangent space of $\mathcal{N}$ at the origin, and we identify this tangent space with $\mathcal{N}$ itself. The operators DH, DH_i of the linearized constraints are defined by the "chain rule,"

$$\delta H = DH[\delta g,\delta\pi] = DH[h,\omega]\,, \tag{19a}$$

$$\delta H_i = DH_i[\delta g,\delta\pi] = DH_i[h,\omega]\,. \tag{19b}$$

We note that the equations of time evolution are a particular, linear map from $\mathcal{N}$ to $T(T^*\mathcal{M})$ in which time plays the role of the parameter λ. We call this linear map E_g,E_π. If $N,N_i\varepsilon\mathcal{N}$ are the usual lapse and shift functions, we have

$$\delta g = dg/dt = E_g(g,\pi)\,[N,N_i]\,, \tag{20a}$$

$$\delta\pi = d\pi/dt = E_\pi(g,\pi)\,[N,N_i]\,. \tag{20b}$$

(For example, $E_g(g,\pi)\,[N,N_i] = 2(\pi_{ij} - \frac{1}{2}g_{ij}\,\pi^k{}_k)\,N + N_{i\,|\,j} + N_{j\,|\,i}$, but we shall not need explicit expressions.)

Next we introduce the natural Hilbert space scalar products in the spaces $T(T^*\mathcal{M})$ and $\mathcal{N}$:

$$(h_1,\omega_1 \mid h_2,\omega_2) \equiv \int(h_{1ij}h_2{}^{ij} + \omega_{1ij}\omega_2{}^{ij})dV \tag{21a}$$

$$(N,N_i \mid H,H_i) \equiv \int(NH + N_iH^i)dV\,. \tag{21b}$$

We note that if the elements N,N_i and H,H_i of $\mathcal{N}$ have their usual meaning of lapse and shift functions (and Hamiltonian and momentum constraints), then the scalar product in (21*b*) is just the total Hamiltonian $\mathcal{H}$. Therefore, we can find DH, DH^i by varying (21*b*) and using the relation of $\mathcal{H}$ to time displacements. In the variation, g and π are functions of λ, but N,N_i are independent of λ:

$$\delta\mathcal{H} = \delta(N,N_i \mid H,H_i) = (N,N_i \mid DH[h,\omega],DH^i[h,\omega]) \tag{22a}$$

$$= \int\left(-\frac{d\pi_{ij}}{dt}\,\delta g^{ij} + \frac{dg_{ij}}{dt}\,\delta\pi^{ij}\right)dV = (-d\pi/dt,\, dg/dt \mid h,\omega)$$

$$= (-E_\pi[N,N_i],\, E_g[N,N_i] \mid h,\omega)\,. \tag{22b}$$

Because N,N_i,h,ω are arbitrary, these equations specify DH, DH^i in terms of E_π, E_g, namely as adjoint operators of each other:

$$(DH,DH^i)^* = (-E_\pi,E_g)\,. \tag{22c}$$

Characterization of Stable and Unstable Spaces

We have seen from the second version of the implicit function theorem that for stability, (DH,DH^i) must be surjective (i.e., its image [of $T(T^*\mathcal{M})$] must fill the entire space $\mathcal{N}$). Since (DH,DH^i) is a linear operator, its image in any case is a linear space. If it does not fill all of $\mathcal{N}$, there are vectors in $\mathcal{N}$ normal to the image, that is, some (N,N_i) for which the scalar product (22*a*) vanishes for *all* h,ω. Equation (22*b*) shows that such N,N_i satisfy $E_g(N,N_i) = E_\pi(N,N_i) = 0$; that is, they specify deformations of the initial surface for which neither g nor π change. But (g,π), as initial values, in turn determine the entire spacetime geometry. Therefore, the deformations N, N_i that leave these geometric objects unchanged must be "initial values" of spacetime isometries, and so give rise to *Killing vectors*. If the spacetime has no symmetries, hence no Killing vectors, then DF is surjective, and the solution is stable (Figure 3.6).

Conversely, if there are Killing vectors, the solution is unstable. We show this by exhibiting second order conditions that the linearized solutions must satisfy in order to be tangent to an exact solution. As before, let $g(\lambda), \pi(\lambda)$ be a curve of exact solutions of the constraints, $F(g(\lambda), \pi(\lambda)) = (H(g,\pi), H_i(g,\pi)) = 0$. The first order (linearized) constraints are obtained by differentiating with respect to λ ($\delta = d/d\lambda$),

$$DF(\delta g, \delta\pi) = 0\,, \tag{23}$$

and setting $\lambda = 0$. Additional conditions may arise as integrability conditions of the second (or higher) order constraints. These are obtained by differentiating (23) once again:

$$D^2F(\delta g, \delta\pi; \delta g, \delta\pi) + DF(\delta^2 g, \delta^2\pi) = 0 \tag{24}$$

and setting $\lambda = 0$. Hence $D^2F(\ \ ;\ \ )$ is a bilinear operator corresponding to the "quadratic" terms in F.

Now suppose there is a spacetime Killing vector, and let N, N_i be its components with respect to the initial surface. By reading equation (22) backwards we then find

$$0 = (N, N_i \mid DH(\delta g, \delta\pi), DH_i(\delta g, \delta\pi)) = (N, N_i \mid DF(\delta g, \delta\pi)) \tag{25}$$

for *all* $\delta g, \delta\pi$, whether they satisfy the linearized constraints or not. Hence, in particular, we may insert into (25) the solutions $\delta^2 g, \delta^2\pi$ of the second order constraint (24), so that we find

$$\begin{aligned} 0 &= -\,(N, N_i \mid DF(\delta^2 g, \delta^2\pi)) = (N, N_i \mid D^2F(\delta g, \delta\pi; \delta g, \delta\pi)) \\ &= \int [N\, D^2H(\delta g, \delta\pi; \delta g, \delta\pi) + N_i\, D^2H^i(\delta g, \delta\pi; \delta g, \delta\pi))dV\,. \end{aligned} \tag{26}$$

This equation is the required second order condition obtained from the Killing vector components N, N_i. One can verify that this condition is automatically preserved by the evolution equation. Thus we obtain exactly one quadratic condition for each Killing vector. These conditions are nontrivial and therefore imply genuine instability, as is shown explicitly by the following example.

Example of Second Order Conditions:
Perturbations of the Flat Three-dimensional Torus

Let us fix the coordinates to order $n = 0$ by requiring $g_{ij} = \delta_{ij}$, $\pi_{ij} = 0$ and to order $n = 1$ (gauge condition) by requiring $\delta g^{ij}{}_{,j} = 0$ (transverse gauge)

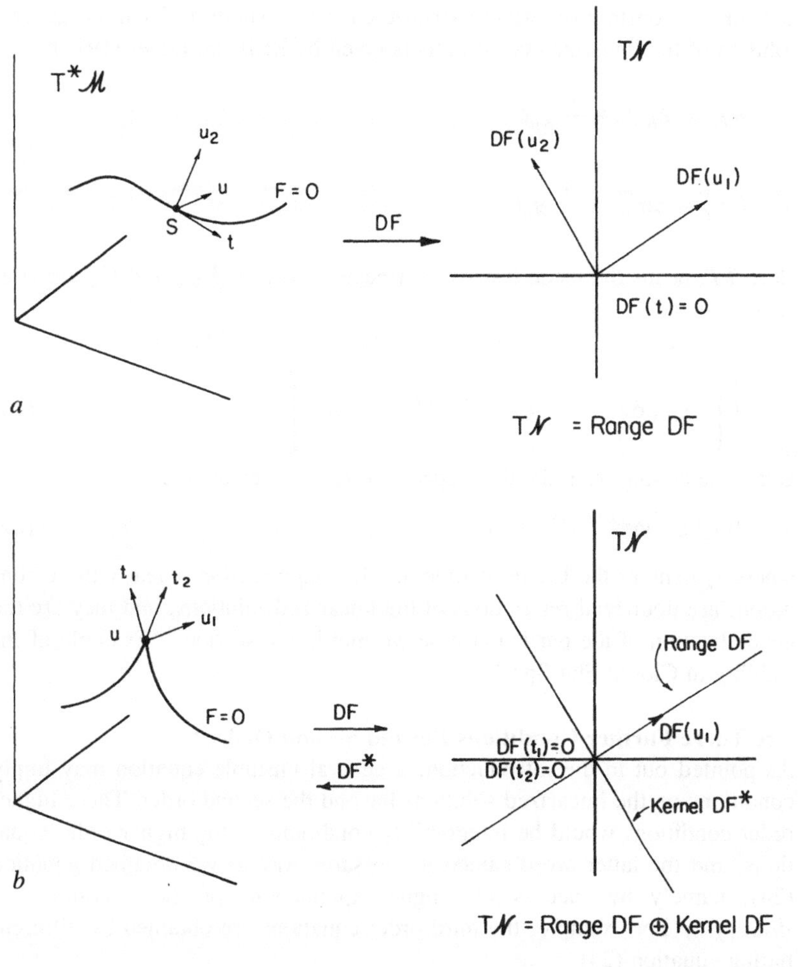

Figure 3.6. Instability and Killing vectors. This issue is summed up in a nutshell form by the equation ("Fredholm alternative for elliptic operators") $T\mathcal{N}$ = Range $(DF) \oplus$ Kernel (DF^*). Here the F's are the constraints; they map $(g,\pi) \varepsilon T^*\mathcal{M}$ to a point of $\mathcal{N}$. The linearization of F, DF, is an elliptic operator. (a) Solutions of the constraint equations $F = (0, 0)$ are represented by a curve in a large dimensional space. The tangent t to the curve at a solution u represents tangents to the solution space and so lies in the Kernel of DF. To decide about stability at u, we investigate whether the image of DF acting on $T_u{}^*\mathcal{M}$ is all of $T\mathcal{N}$. If so, it follows from the implicit function theorem that the solution u is stable. (b) We have an instability symbolized by the "corner" in the solution space at u. Here DF is not surjective. Hence Range $(DF) \neq T\mathcal{N}$ and so Kernel $(DF^*) \neq 0$. But from equation (22b) of the text, Kernel (DF^*) consists of Killing vectors. Hence, in the unstable case there must be Killing vectors. As in the text, $\mathcal{N}$ and $T\mathcal{N}$ may be identified. (Figure provided by Murray Cantor.)

and $\delta\pi'_i$ = constant (constant extrinsic curvature slicing). Then the general solution of the linearized constraints is given by Brill and Deser (1973):

$$\begin{aligned} \delta g_{ij} &= \delta g^{TT}_{ij} + \frac{1}{3}\, g_{ij} C_g \\ \delta \pi_{ij} &= \delta \pi^{TT}_{ij} + \frac{1}{3}\, g_{ij} C_\pi \,. \end{aligned} \tag{27}$$

Here TT means the usual transverse traceless part, and C_g and C_π are constant in space.

The second order conditions (26) due to the timelike Killing vector are

$$\int \left[\delta g^{TT}_{ij,k}\, \delta g^{ij,kTT} + \delta \pi^{TT}_{ij}\, \delta \pi^{ijTT} - \frac{1}{6} C^2_\pi \right] d^3x = 0\,, \tag{28}$$

and those arising from the three spacelike Killing vectors are

$$\int (\pounds_k \delta g^{TT}_{ij}) \delta \pi^{ijTT}\, d^3x = 0\,, \tag{29}$$

where $\pounds_k$ denotes the Lie derivative in a Killing direction. Clearly these conditions are nontrivial restrictions of the linearized solutions, and they are the generalization of the particular case we met in the section, "Example of Instability in Closed Flat Space."

Are There Further Conditions Beyond Second Order?

As pointed out in the introduction, a general unstable equation may imply conditions on the linearized solutions beyond the second order. These higher order conditions would be integrability conditions of the higher order equations; and the latter are obtained in the same way as we obtained equation (24), namely by successively higher applications of the operator $\delta = d/d\lambda\,|_{\lambda=0}$. For example, the third order equations are obtained by differentiating equation (24):

$$D^3F(\delta g,\delta\pi)^3 + 3D^2F(\delta^2 g,\delta^2\pi;\delta g,\delta\pi) + DF(\delta^3 g,\delta^3\pi) = 0\,. \tag{30}$$

The conditions that this equation be integrable for $\delta^3 g,\delta^3\pi$ follow from the same scheme as before

$$(N,N_i\,|\,DF(\delta^3 g,\delta^3\pi)) = 0\,. \tag{31}$$

However, here it is not immediately clear whether, after substituting from equation (30), this requirement reduces to some conditions on the $\delta g,\delta\pi$ or whether equation (31) can always be satisfied by appropriate choice of solutions $\delta^2 g,\delta^2\pi$ of equation (24). The latter would be the normally expected generic situation, in which there are no third (or higher) order conditions and hence no third or higher sources of instability.

V. Moncrief (personal communication, 1977) noted that in general relativity some of the constraints, namely $\pi_i^j{}_{|j} = 0$ written in this covariant form, contain no terms higher than second order in the g,π. Hence equation (30) and all higher order equations vanish identically, and so do the corresponding integrability conditions. Therefore, we cannot obtain conditions beyond second order from the momentum constraints. But whatever the complete set of conditions, we expect them not essentially to distinguish between the Hamiltonian and momentum constraints (see, e.g., equation (26) and problem 11 below). It is therefore reasonable to expect that no conditions arise beyond second order from any of the constraints. Indeed it appears that equation (24) is sufficient (as well as necessary) to characterize all the linearized solutions that approximate exact solutions if there are Killing vectors; and if there are no Killing vectors, we have stability, so that all linearized solutions have exact solutions corresponding to them (V. Moncrief, personal communication, 1977; J. Marsden, personal communication, 1977).

4. SUMMARY

In broad outline the linearization stability properties of Einstein's equations are today well understood. Asymptotically flat spacetimes are linearization stable. Spatially closed spacetimes are stable if they do not admit any Killing vectors. If the spacetime does have Killing vectors, these generate second order conditions, which distinguish those solutions of the linearized equations that correspond to exact solutions.

Our discussion has been confined to the vacuum Einstein equations. A number of cases with matter sources have been investigated (Arms, 1977; d'Eath, 1976) and give results quite similar to the sourceless case.

Thus one has today a fairly complete answer to the original question about the reliability of the linearized approximation. We have recognized it as just one of the local properties of the solution "manifold" within the space of all metrics. Of course, many further questions remain about the local, and eventually about global, properties of this manifold and about the corresponding subspace of the space of all geometries. I hope that the above treatment has given a glimpse of the sort of methods that can be useful in investigating these further questions.

ACKNOWLEDGMENTS

I thank the University of Texas and colleagues there for their hospitality during my sabbatical when this work was in progress. I also thank M. Cantor and

J. Isenberg for reading the manuscript and for valuable suggestions. This work was supported in part by the National Science Foundation.

EXERCISES AND PROBLEMS

1. To show that linearization stability has nothing directly to do with the usual Liapunov stability (for the latter see, for example, Hale, 1969), give an example that is stable in one sense but unstable in the other. (Hint: All linear equations are linearization stable.)

2. Consider the minimal surface equation, that is, equation (1), in the following two cases: (a) surface in $T^2 \times R$, with the torus T^2 treated by imposing periodic boundary conditions on $f(x,y)$; (b) surface in $D^2 \times R$, with f constrained to vanish on the circular boundary of the disk D^2. Show in both cases that the minimal surface equation is linearization stable. (This demonstration can be done using either the results from the section "Example of an Unstable Equation" or the theorem from the section "Second Version of the Theorem.")

3. Show that the scalar wave equation resulting from the "ϕ^4" Lagrangian is linearization stable.

4. Linearize the geodesic equation, regarded as a second order equation, about a geodesic with tangent T. The variation can be regarded as a vector field W along the geodesic. The result is the Jacobi equation. (See, for example, Hicks, 1965, p. 144.) Show that the geodesic equation is linearization stable, for example, in some simple case. Compare and contrast with the minimal surface equation discussed in the section "Example of an Unstable Equation" and in problem 7.

5. In the section "Geometrical Formulation of Linearization Stability" the equations $x^4 + y^6 = 0$ and $(x - y)^2 = 0$ are mentioned as counterexamples to some simple ideas about the unstable case. Compare the exact solutions of these equations with the solution of an approximation to linear and higher orders of these equations, and verify the statements made in the text.

6. Use the "modified Newton's method" (Figure 3.4) to find the zero of $F(x,y) = x - y^2$ at $x = 2$. (So this will approximate $y = \sqrt{2}$.) Start with $x_0 = y_0 = 1$. Do a few iterations "by hand." (Answers: $y_1 = 3/2$; $y_2 = 11/8$; $y_3 = 183/128 \ldots$.) By evaluating the "contraction factor" k, estimate the number of iterations necessary for an accuracy of 1%. (Answer: $1/4 \leq k \leq 1/2$, asymptotically $k \to \sqrt{2} - 1$. Because $[1/2]^7 < 0.01$, seven iterations would be plenty.)

7. Carry through an iteration as in the section "A 'One-dimensional' Example" for the minimal surface equation (1). The "unperturbed" solution is $f = 0$. For the first order solution of equation (2) take $g = x^2 - y^2$. Show that the iteration fails in third order.

8. The converse of the theorem of the section "Second Version of the Theorem" does not hold: an equation $F(Z) = 0$ can be linearization stable at Z_0 even if $DF(Z_0)$ is not surjective. To convince yourself, work out the case $F: R \to R^2$ by $x \to (x,x^2)$.

9. Use the explicit expressions for E_g, E_π, substitute into equation (22*b*), and integrate by parts to remove the derivatives on the N,N_i to get a result in the form of equation (22*a*). Hence read off the linearized constraints DH, DH^i (about any solu-

tion, not necessarily flat). Verify agreement with a direct variation of the constraints H, H^i. For the integrations by part in this problem it is easiest to regard π^{ij} as densities, hence not use the expression for E_g as given after equation (20*b*), but instead use the standard expressions found, for example, in equations (21.114) and (21.115) of Misner, Thorne, and Wheeler (1973). For the solution, see Moncrief (1976).

10. Asymptotically flat spaces with Killing vectors do not show the linearization instability derived for closed spaces in the section "Characterization of Stable and Unstable Spaces"; the argument fails because equations (22*a*,*b*,*c*) are no longer valid. Show that in asymptotically flat spaces, equations (22*a*,*b*,*c*) need to be supplemented by terms of the type $N_\infty \delta m$, where N_∞ is the asymptotic value of the lapse and δm is the variation of the total mass, and by similar terms in the variation of the total linear and angular momentum. Hence, show that equation (26) is no longer a quadratic condition on the linearized solution, but instead specifies the second order change in the system's total mass (or momentum or angular momentum, depending on which Killing vector is present).

11. Show that the second order condition, equation (26), can be written in the spacetime form

$$\int k^\mu D^2 G_\mu{}^\nu(\delta \overset{4}{g}, \delta \overset{4}{g}) n_\nu d^3S = 0 ,$$

where k = Killing vector, G = Einstein tensor, and $\delta \overset{4}{g}$ = variation of spacetime metric. Also show that the variation of the Bianchi identities guarantees that this equation holds on all spacelike surfaces if it holds on one. (See Moncrief, 1976.)

12. The second order conditions show possible singularities in the solution subspace of the phase space $T^*\mathcal{M}$, where $\mathcal{M}$ = Riem (M). The group Diff(M) of diffeomorphisms of M, extended in the standard way to $T^*\mathcal{M}$, has singularities at the same points, namely whenever a three-space Killing vector exists. (See Fischer, 1970, p. 303.) Form the quotient "super phase space," $T^*\mathcal{M}$/Diff(M). Use the explicit expressions of the section "Example of Second Order Conditions" to find out whether, for the case discussed there, the singularities "cancel out," making the solution subspace of super phase space regular at this point (cf. Fischer and Marsden, 1975, p. 58).

13. Discuss the linearization stability of Einstein gravity coupled to a scalar field or of the Brans-Dicke theory (see Misner, Thorne, and Wheeler, 1973, p. 1070). In the compact case, show that there is instability if there is a combined Killing vector for the metric and the scalar field. (For the analogous problem with electromagnetism, see Arms, 1977.)

NOTES

1. See, for example, H. B. Lawson, Jr., "Minimal Varieties," in *Differential Geometry* (Proceedings of Symposia in Pure Mathematics, vol. V, part 1, Amer. Math Soc., Providence, Rhode Island, 1975), p. 143, theorem 2.14.

2. It is interesting to note that this equation occurs in general relativity as a description of the Kasner universes within the space of anisotropic cosmologies described by metrics (with $\beta_\pm$ functions of t only),

$$ds^2 = -dt^2 + e^{2\Omega}(e^{2\beta_+ + 2\sqrt{3}\beta_-}\,dx^2 + e^{2\beta_+ - 2\sqrt{3}\beta_-}\,dy^2 + e^{-4\beta_+}\,dz^2)\,.$$

We let $X = d\beta_+/dt$, $Y = d\beta_-/dt$, $Z = d\Omega/dt$; then the Einstein equation $G_{00} = 0$ becomes $X^2 + Y^2 - Z^2 = 0$. The origin then represents flat spacetime; the upper cone, the expanding Kasner universes; and the lower cone, the contracting Kasner universes. For more detail, see Ryan and Shepley (1975).

3. One might expect that at the first iteration step, the effective total unaccounted energy, $1/16\pi \int [-R(g)+\overline{\pi}_{ij}\,\overline{\pi}^{ij}]d^3x$ should equal the physical total energy of the system. However, the integral of equation (11) shows that the latter is given by a "red-shifted" version of the former, namely $1/16\pi \int [-R(g)\psi + \overline{\pi}_{ij}\,\overline{\pi}^{ij}\,\psi^{-7}]\sqrt{g}\,d^3x$.

4. The Taub universe (Ryan and Shepley, 1975) does have instabilities in the momentum constraints due to its spacelike symmetries. (See "Applications of the Implicit Function Theorem" section.)

5. Here and in the following we regard all quantities—in particular, H, H_i, π_{ij}—as tensors and not as densities.

REFERENCES

Arms, J. M., "Linearization Stability of the Einstein-Maxwell System," *J. Math. Phys.* **18**, 830–833 (1977).

Brill, D. R., and Deser, S., "Instability of Closed Spaces in General Relativity," *Commun. Math. Phys.* **32**, 291–304 (1973).

Cantor, M., "The Existence of Non-trivial Asymptotically Flat Initial Data for Vacuum Spacetimes," *Commun. Math. Phys.* **57**, 83–96 (1977).

Choquet-Bruhat, Y., "*Stabilité par Linearization*," conference at Vendome, France (May 1977), and "Space of Asymptotically Euclidean Initial Data," GR8 conference, Waterloo, Canada (August 1977).

Choquet-Bruhat, Y., and Deser, S., "On the Stability of Flat Space," *Ann. Phys. (USA)* **81**, 165–178 (1973).

Choquet-Bruhat, Y., DeWitt-Morette, C., Dillard-Bleick, M., *Analysis, Manifolds and Physics* (North-Holland Publishing Co., 1977), chapter 2.

Choquet-Bruhat, Y., Fischer, A. E., and Marsden, J. E., "Maximal Hypersurfaces and Positivity of Mass," in *Isolated Gravitating Systems in General Relativity* (edited by J. Ehlers, North-Holland Publishing Co., Amsterdam, 1979), pp. 396–456.

d'Eath, P., "On the Existence of Perturbed Robertson-Walker Universes," *Ann. Phys. (USA)* **98**, 237–263 (1976).

Fischer, A. E., "The Theory of Superspace," in *Relativity* (edited by M. Carmeli, S. I. Fickler, and L. Witten, Plenum Press, New York, 1970), pp. 303–357.

Fischer, A. E., and Marsden, J. E., "Linearization Stability of Nonlinear Partial Differential Equations," *Proceedings of Symposia in Pure Mathematics* **27**, 219 (1975).

Hale, J. K., *Ordinary Differential Equations* (Wiley-Interscience, New York, 1969), chapter 1.

Hawking, S. W., and Ellis, G. F. R., *The Large Scale Structure of Space-time* (Cambridge Univ. Press, Cambridge, 1973), section 7.6.

Hicks, N. J., *Notes on Differential Geometry* (Van Nostrand Co., Princeton, New Jersey, 1965).

Misner, C. W., Thorne, K. S., and Wheeler, J. A., *Gravitation* (W. H. Freeman and Co., San Francisco, 1973), chapters 18 and 21.

Moncrief, V., "Spacetime Symmetries and Linearization Stability of the Einstein Equations: I," *J. Math. Phys.* **16**, 493–498 (1975).

———, "Spacetime Symmetries and Linearization Stability of the Einstein Equations: II," *J. Math. Phys.* **17**, 1893–1902 (1976).

Ō Murchadha, N., and York, J., "Initial-value Problem of General Relativity: I. General Formulation and Physical Interpretation," *Phys. Rev. D* **10**, 428–436 (1974*a*).

———, "Initial-value Problem of General Relativity: II. Stability of Solutions of the Initial-value Equations," *Phys. Rev. D* **10**, 437–446 (1974*b*).

Ryan, M. P., and Shepley, L. C., *Homogeneous Relativistic Cosmologies* (Princeton Univ. Press, Princeton, N.J., 1975).

4. Nonlinear Model Field Theories Based on Harmonic Mappings

CHARLES W. MISNER

The aim of this paper is to discuss something of what is known about the action functional (Misner, 1978)

$$I = \int G_{AB}(\phi)\left(\frac{\partial\phi^A}{\partial x^\mu}\right)\left(\frac{\partial\phi^B}{\partial x^\nu}\right) g^{\mu\nu}\sqrt{g}\, d^4x \tag{0.1}$$

and its relatives, where ∂ is replaced by a gauge covariant derivative. In particular, I suggest that theories using an action of this form—that have so far appeared only "accidentally" in physics—should be more exhaustively surveyed for possible applications.

The special feature of this action functional is that it contains *two* metrics on two different spaces. One space is spacetime, where we will most commonly adopt the flat Minkowski metric $g_{\mu\nu}(x) = \eta_{\mu\nu}$ (Misner, Thorne, and Wheeler, 1973). The other space, to give theories that are interesting and novel, should be curved. Then linear combinations of ϕ's are not meaningful, so no linear equation for the ϕ field can be written; therefore, the field equations following from this action functional will appear natural although nonlinear. The functions $\phi^A(x^\mu)$ that extremize I are called harmonic maps (Eells and Sampson, 1964; Eells and Lemaire, 1978; Fuller, 1954).

This class of field theories, therefore, represents a class (different from the family of Yang-Mills theories [Yang and Mills, 1954; Coleman, 1975; Moriyasu, 1978]) in which self-coupling necessarily exists in a form specified by internal symmetry.

This paper has three parts: history and motivation from general relativity, examples related to particle physics, and open questions. This paper defines an area for research; it is not a review of extensive completed work.

1. HISTORY AND MOTIVATION OF THE USE OF HARMONIC MAPS IN PHYSICS

The first example of the use of harmonic mappings in physics that I encountered was in the solution of the Einstein equations of general relativity in a simplified situation with axial symmetry. This was work I did with Richard Matzner (Matzner and Misner, 1967), but we did not then know that what we were dealing with was a harmonic map. We had obtained field equations for two functions α and β that fixed the spacetime metric for our problem, and we recognized that they were derivable from a variational principle with a very geometrical character. The Lagrangian was

$$\pounds = (\nabla\alpha)^2 - \cosh^2\alpha\ (\nabla\beta)^2 . \tag{1.1}$$

Here the ∇ operators are ordinary flat space gradients, and the implied inner product is the three-dimensional Euclidean one. But another metric shows itself in this Lagrangian, an unexpected metric defined (by the Einstein equations as here reduced to a special case) on $\alpha\beta$ space as

$$ds'^2 = d\alpha^2 - \cosh^2\alpha\ d\beta^2 . \tag{1.2}$$

Thus the dynamics of the problem we were attacking had assigned this specific geometry to the manifold M' of possible $\alpha\beta$ field values, a geometry entirely different from the various curved geometries that the solutions of these field equations would give to spacetime M.

I was very intrigued by this unexpected geometry we had found buried in the Einstein equations and also by the possibility that a Lagrangian with so much geometry governing its construction would be accompanied by a powerful theory describing the nature of the solutions of its field equations. Continued inquiry eventually led me (with the help of S. Smale, personal communication, 1969) to the fundamental paper by Eells and Sampson (1964) that describes this class of nonlinear partial differential equations that are known as harmonic maps. (The important earlier publication of Fuller, 1954, is shorter and outlines many major points.) The defining features of a harmonic map are its variational principle, which I have already given as equation (0.1), and the fact that what the resulting field equations control is a mapping from one Riemannian manifold M with metric ds^2 to another Riemannian manifold M' with its own metric ds'^2. This is summarized in Figure 4.1.

About the time I learned of the Eells-Sampson paper on harmonic maps, Yavuz Nutku was at the University of Maryland developing other ideas of "the geometry of dynamics" (Nutku, 1974). He was as intrigued as I was with the way geometry determined dynamics in harmonic maps and proceeded to find other applications of them in general relativity. His most recent such result is an application to interacting plane gravitational waves (Nutku and Halil, 1977).

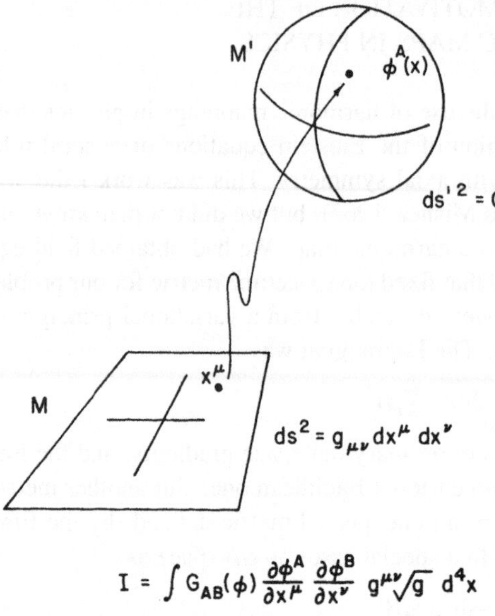

Figure 4.1. The action defining harmonic maps contains the metric $g_{\mu\nu}$ of spacetime M, but also defines a typically nonflat metric G_{AB} on the space of field values, M'. Physically, a harmonic map (which renders the action stationary) assigns a value to the field ϕ^A for every point x^μ in spacetime. Nonlinearities arise because of the appearance of $G_{AB}(\phi)$ in the action.

The application of harmonic mappings in physics that may be most widely known is the nonlinear σ model. This model theory in particle physics is over a decade old (Gell-Mann and Lévy, 1960). Duff and Isham (1977) and Appelquist and Bernard (1980, 1981) give recent results on it and use a formalism that displays clearly its character as an harmonic mapping theory; but perhaps only Nutku (1974) has noted in print that it is an example of the class of equations called harmonic maps defined by Fuller (1954) and further studied by Eells and Sampson (1964).

In the preceding paragraphs I have used "harmonic map" only to describe field equations that could not be readily identified as belonging to some more familiar class. But it is also valuable to note that many very familiar equations are special cases of the harmonic mapping equation; this familiarity provides harmonic mapping theories with some kind of correspondence principle that gives them roots in better established theories and therefore makes the study of them appear more plausibly fruitful. Eells and Sampson (1964) noted a number of such mathematical special cases of harmonic maps. I will list those that are of some physical interest.

The special case from which harmonic maps get their name is that of harmonic functions, that is, real valued functions satisfying the Laplace equation. This case occurs when the manifold M' of field values is one-dimensional and therefore flat (with $G_{AB} = 1$ for an appropriate scaling of the field ϕ). In physics this will lead to a wave equation when the spacetime metric $g_{\mu\nu}$ in equation (0.1) has Minkowski signature. Thus, the usual linear wave equation is a special case of the harmonic mapping equation.

But in addition to generalizing the wave equation, the harmonic mapping equation also generalizes another equation that is familiar at least in gravitational physics, namely the geodesic equation. If the domain space M is taken to be one-dimensional (and thus flat with $g_{\mu\nu} = 1$ natural), equation (0.1) states the variational integral from which the geodesic equation results (Misner, Thorne, and Wheeler, 1973). The geodesic equation is, of course, normally a nonlinear, although ordinary, differential equation. *In the harmonic mapping equation, then, we can expect the nonlinearities of the geodesic equation compatibly combined with the causality and wave propagation features of wave equations.*

Further mathematical examples of harmonic maps may also be familiar to many physicists (see Eells and Sampson, 1964, using Helgason, 1978, for definitions). Isometries and coverings of Riemannian manifolds $M \rightarrow M'$ are harmonic maps; so also are minimal immersions $M \rightarrow M'$, which have appeared in useful coordinate conditions for numerical computations in general relativity ("maximal time slices") (DeWitt et al., 1973) as well as in the theory of soap films (Almgren, 1966). All homomorphisms $G \rightarrow G'$ of simple Lie groups are also harmonic maps, which might point to some relationship between the harmonic mapping equation and gauge theories. Similarly, all holomorphic maps of Kähler manifolds—which occur, for instance, in twistor theory (Penrose and MacCallum, 1973)—are also harmonic maps. Thus harmonic maps are mathematical structures close to several that are known to be good models of parts of the physical world and to some others that are the subject of current investigations.

Some relationship to known physics—as in harmonic mappings' relationship to the linear wave equation—is a valuable prerequisite to trying out a bit of unfamiliar mathematics, but certainly not adequate motivation for active exploration. As active motivation for studying harmonic mapping theories, I will suggest four items. The first two will receive the most attention in the remainder of this paper. The four are (1) modeling nonlinearities of the Einstein equations, (2) natural symmetry breaking, (3) dynamic structuring of bundles, (4) soluble (?) quantum field models. The first of these points will be addressed immediately; the others will appear in remarks during and after the presentation of several examples in the next part of this paper.

I find harmonic maps intriguing for physical speculation because they can be gathered with Einstein's gravitation theory, the Yang-Mills gauge theories, and theories with local supersymmetries (Teitelboim, 1980) in a limited class of field theories that can be called *naturally nonlinear fields*. These fields are characterized by the properties that no linear field equation is available that respects the adopted symmetries and that a nonlinear field equation is natural and contains only a small number of adjustable coupling constants. Thus one does not introduce interactions among the component fields ad hoc (as by adding higher polynomial terms to a quadratic free-field Lagrangian) but has them enforced by general symmetry principles.

Within this class of theories it is possible to view the Yang-Mills theories and the harmonic mapping theories as models for the apparently more difficult nonlinearities of general relativity. Each, however, models general relativity in a different way. During nearly a decade before their current important uses in unified gauge theories of particle interactions were established, Yang-Mills theories were, in fact, employed fruitfully as models for general relativity by workers developing covariant quantization methods for gravitation (Feynman, 1972; DeWitt, 1967; Faddeev and Popov, 1967; Mandelstam, 1968). But harmonic mapping theories should model quite different aspects of the Einstein gravitational equations than do the Yang-Mills theories and might, therefore, allow quite new progress to be made on the quantization of the gravitational field.

To best see the analogies and distinctions among these three classes of naturally nonlinear fields, it is convenient to write their field equations in a sketchy form:

Gravity

$$g^{\alpha\beta}\frac{\partial^2 g_{\mu\nu}}{\partial x^\alpha\,\partial x^\beta} + g\left(\frac{\partial g}{\partial x}\right)^2 = 0\,,$$

Harmonic maps

$$\frac{\partial}{\partial x^\alpha}\left[G_{AB}(\phi)\frac{\partial\phi^B}{\partial x_\alpha}\right] + \Gamma(\phi)(\partial\phi)^2 = 0\,, \tag{1.3}$$

Yang-Mills

$$\frac{\partial^2 A^i{}_\mu}{\partial x^\alpha\,\partial x_\alpha} + A\,\partial A + A^3 = 0\,.$$

All these equations have in common that they are quasi-linear, that is, the second derivatives appear linearly.

The Yang-Mills and harmonic mapping equations have the further simplification that the second derivatives form a simple flat-space wave operator $\eta^{\alpha\beta}\partial_\alpha\partial_\beta$ (where $\eta^{\alpha\beta}$ is the Minkowski metric). In the lower order terms, however, one sees that the harmonic mapping equation shares with the Einstein equation the property of being quadratic in first derivatives so that the entire equation is homogeneous of degree 2 in the derivatives ∂. The Yang-Mills equation, in contrast, has a different character, reflected in the dimensions of $A^i{}_\mu$ that are naturally taken as $(\text{length})^{-1}$, while $g_{\alpha\beta}$ and the harmonic map ϕ can be naturally taken as dimensionless. This is usually considered a virtue of the Yang-Mills field, because it is related to the renormalizability of its quantum field theory.

But if our interest is focused upon gravitation, this feature serves to point up the idea that harmonic maps may be good models of some of the problems of quantizing general relativity that have not yet been solved by the methods that used Yang-Mills theories as stepping stones. There is no assurance that these deep problems can be solved without seriously changing Einstein's theory, but the suggestion here is that by using harmonic mapping theories as models, some attempts at nonperturbative quantization might be launched. There is more hope of success with harmonic maps than (initially) with general relativity, because harmonic mapping theories are of a large class that contains nonlinear theories much simpler than general relativity.

In comparing the field equations above, there is a further apparent similarity between the harmonic mapping theories and general relativity that may make harmonic maps good models for general relativity theory. This is the fact that, as written above, both contain nonlinearities in nonconstant coefficients of the second derivative terms, while in the Yang-Mills equation the second derivative terms is entirely linear. In the classical theories this distinction is specious, since the harmonic mapping equations can be recast, by multiplying them by the inverse matrix G^{AB}, to a form with the leading term $\partial^2\phi/\partial x^\alpha \partial x_\alpha$. In the quantum theory, however, such a reformulation is likely to be impossible as a result of operatór ordering problems. Consequently, this particular nonlinearity of the Einstein equations would be present in a significant but mild way in harmonic mapping theories, which then become valuable training grounds for the quantum treatment of this type of nonlinearity.

2. EXAMPLES RELATED TO PARTICLE PHYSICS

An Elementary Example

As further introduction to harmonic maps, let us consider some examples. The first is geometrically simple but also related to the Sine-Gordon equation that has been of interest in several areas of physics (Scott, 1969; Jackiw, 1977). This first example considers a map from the Euclidean or Minkowski n-plane R^n into the sphere S^2, written $\phi: R^n \rightarrow S^2$. The field equation will result from a variational principle, and for the map ϕ to be a harmonic map, the action integral to be varied is

$$I = \frac{1}{2} \int [(\nabla\Theta)^2 + \sin^2\Theta\, (\nabla\Phi)^2] d^n x . \tag{2.1}$$

The field equations that result are

$$\Delta\Theta + \sin\Theta \cos\Theta\, (\nabla\Phi)^2 = 0 , \tag{2.2a}$$

$$\sin\Theta\, \Delta\Phi - 2 \cos\Theta\, (\nabla\Theta)\cdot(\nabla\Phi) = 0 . \tag{2.2b}$$

Here Θ and Φ are the usual angular coordinates on a sphere, and $\Delta = -\partial_\alpha \partial^\alpha$ is the wave operator or Laplacian, depending on the signature chosen for the metric on R^n.

The sphere S^2 here is the space of field values that might perhaps be thought of as the direction of an internal isospin-like degree of freedom in some classical limit. Or, as in the nonlinear σ model, the Θ and Φ fields may be considered to be the two independent fields among three scalar fields X,Y,Z constrained by an identity $X^2 + Y^2 + Z^2 = 1$. The field equations (2.2a,b) are a pair of coupled nonlinear wave equations for these two scalar fields.

The interpretation of $\Theta\Phi$ as coordinates on a sphere becomes important when one asks whether a proposed solution of these equations is singular or not. Thus, the behavior of Θ and Φ must be such that $Z = \cos\Theta$, $X = \sin\Theta \cos\Phi$, and $Y = \sin\Theta \sin\Phi$ are differentiable functions of the spacetime coordinates $x^1, \ldots, x^n$ on R^n; then

$$\phi:(x^1, \ldots, x^n) \rightarrow (\Theta,\Phi) \text{ and } (x^1, \ldots, x^n) \rightarrow (X,Y)$$

are two coordinate representations of the harmonic map $\phi: R^n \rightarrow S^2$.

For a further specialization of this example, consider the simple case $n = 2$, so that the map in question is $\phi: R^2 \rightarrow S^2$. Particular solutions of the field equations may then be sought in the form

$$\Theta = \Theta(r) , \qquad \Phi = \Phi(\psi) ,$$

where r and ψ are polar coordinates on R^2 and a Euclidean metric for R^2 is assumed. Since $\nabla\Theta$ and $\nabla\Phi$ are orthogonal under these assumptions, equation (2.2b) becomes $\Delta\Phi = 0$, and a particular solution is $\Phi = m\psi$. Now equation (2.2a) derives from the simplified variational principle

$$\begin{aligned} I &= \pi \int_0^\infty r\, dr \left[\left(\frac{d\Theta}{dr} \right)^2 + m^2 \frac{\sin^2\Theta}{r^2} \right] \\ &= m\pi \int_{-\infty}^{\infty} d\rho \left[\left(\frac{d\Theta}{d\rho} \right)^2 + \sin^2\Theta \right] . \end{aligned} \tag{2.3}$$

In the second form the independent variable has been changed from r to $\rho = ln(r/R)^m$ so that the resulting differential equation becomes that for a simple Newtonian particle in a potential $V = -\sin^2\Theta$, as suggested by Figure 4.2.

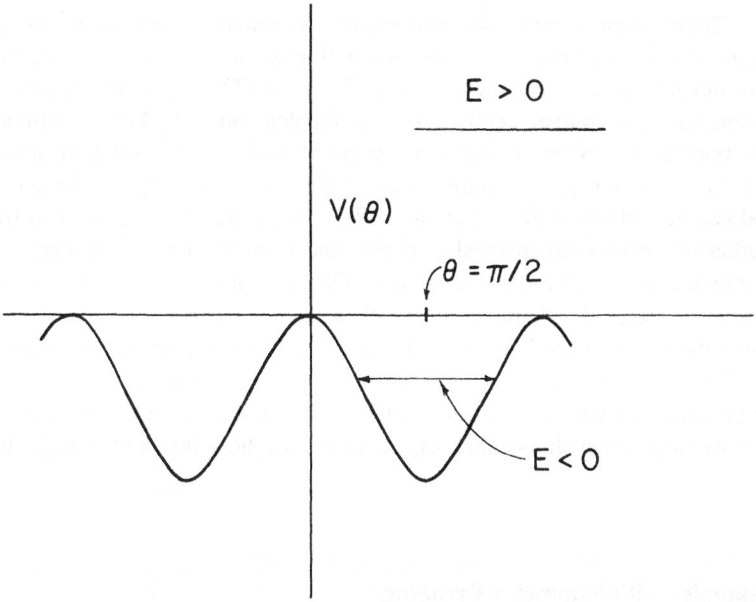

Figure 4.2. The action (2.3), which describes classical isospin as a function of position, can be recast in a form resembling that for classical one-dimensional motion with coordinate θ, under a potential $-\sin^2\theta$. The parameter analogous to time is $\rho = \rho n(r/R)^m$ (R constant), where r is the usual radial coordinate. If the $E > 0$ motion is taken, then clearly $dE/d\rho \geq E^{1/2}$, and there is no regular limit in θ, ϕ behavior as $r \to 0$. By a similar argument, $E < 0$ is excluded; then θ has a limited range as ρ varies. These solutions oscillate rapidly as $r \to 0$. Only the $E = 0$ solutions are acceptable. They smoothly $[\theta = 2\arctan(r/R)^m]$ approach $\theta = 0$ as $r = 0$. (The $\theta =$ constant solutions are acceptable only for $m = 0$, as discussed in the text.)

Experience from elementary mechanics then serves us qualitatively as an analog computer to survey the solutions of this differential equation. We see that for positive "energy" $E = \Theta'^2 + V$, Θ will be a monotonically increasing function of ρ with $d\Theta/d\rho \geq \sqrt{E}$ so that the region of R^2 described by $-\infty < \rho < 0$ or $0 < r < R$ gets mapped infinitely many times over the sphere S^2 (i.e., Θ covers the entire range $-\infty$ to 0). As $r \to 0$, there is therefore no limiting value (as a point in S^2) for the map ϕ or its representatives $\Theta\Phi$, and these solutions are rejected as singular at the origin of R^2. By similar arguments, one rejects the negative "energy" solutions since $\Theta(r)$ has a nonzero range of oscillation in every neighborhood of $r = 0$ and is again not continuous. In the solution with $\Theta = \pi/2 =$ constant, one sees that for $m \neq 0$, the functions $X = \cos m\psi$ and $Y = \sin m\psi$ do not approach a limit as coordinates on S^2 as $r \to 0$ in R^2.

There then remain as nonsingular harmonic maps $\phi : R^2 \to S^2$: $(r,\psi) \to (\Theta,\Phi)$ only the $m = 0$ solution with constant $\Theta\Phi$, and for each nonzero integer m the solution with "energy" $E = 0$. (The requirement that m be an integer is, of course, a consequence of the demand that XYZ be continuous functions on R^2, where ψ and $\psi + 2\pi$ are identified.) This solution satisfies $d\Theta/d\rho = \pm\sin\Theta$, gives a finite value for the action, namely $I = 4\pi \mid m \mid$, and can be written as $\Theta = 2 \arctan(r/R)^m$. These facts are well known from studies of the two-dimensional σ model, and from the Sine-Gordon equation to which equation (2.2a) reduces when $(\nabla\Phi)^2$ is constant. The derivation was repeated here so that familiarity with these theories is not prerequisite to the illustration this example gives of the powerful constraints that geometrical regularity imposes on solutions of the harmonic mapping equation—exemplified here by the $E = 0$ condition on acceptable solutions of the equation for $\Theta(r)$ and the inadmissibility of, for example, the solution $\Theta = \pi/2$, $\Phi = \psi$.

Examples with Symmetry Breaking

Broken symmetry has been an important theme in particle physics for many years now (see, e.g., Bernstein, 1974, and references therein; Coleman, 1975; Taylor, 1976; Faddeev and Slavnov, 1980; or Moriyasu, 1980). Harmonic maps have a deep relationship to some form of symmetry breaking, but whether it is a form that is important or useful for elementary particle theory remains to be investigated. This relationship can be derived from a theorem of Steenrod's concerning fiber bundles. The theorem is given in Steenrod (1951, p. 43): Let H be a closed subgroup of G; a principal G-bundle admits a restriction to a principal H-bundle if and only if the associated G-bundle with fibre G/H has a cross section.

For our present purposes, this theorem says roughly that for any desired breaking of a gauge symmetry, with a gauge group G being reduced to a subgroup H, there is a harmonic mapping field equation available whose solutions can effect the reduction. The relationship is that (at least when H is a compact subgroup of a semisimple Lie group G) the manifold G/H is a curved space with a natural Riemannian metric and can be taken as the manifold M' of field values ϕ for a harmonic mapping field. Solutions of this field equation for ϕ are—provided everything is made gauge covariant—just the cross sections referred to in Steenrod's theorem. For the rest here we shall make do with a few examples rather than attempting to develop a general theory (cf. Misner, 1978).

The set of examples related to symmetry breaking that will be taken up now use an action integral of the form

$$I = I_A + I_\phi + I_\psi, \tag{2.4}$$

where I_A is the Yang-Mills action for a gauge vector field (connection) A_μ, I_ϕ is the harmonic mapping action of equation (0.1) for the symmetry-breaking field, and I_ψ is the action for some set of other gauge covariant fields that we shall not consider explicitly in these examples. They will now, however, let us outline the ideas of gauge covariance that will be used.

Suppose that ψ^a is an isovector field. Then, under a gauge transformation defined by a varying gauge group element $g(x) \varepsilon G$, the ψ field would transform according to the rule

$$\psi \rightarrow g\psi \,. \tag{2.5}$$

For references on these techniques, see DeWitt (1965) or Moriyasu (1978). In order to be able to form gauge covariant derivatives of ψ, one needs to introduce the Yang-Mills field A. One understands that A describes the motion of the basis isovectors to which the index on ψ^a refers, and that the second term $A\psi$ in the formula

$$D\psi = d\psi + A\psi \tag{2.6}$$

for the gauge covariant derivative of ψ arises precisely from the changes in these basis vectors from point to point, just as the first term $d\psi$ arises from changes in the components ψ^a of ψ. (In this equation, a matrix notation is being used for A, so that $A\psi$ stands for $A^a{}_b\psi^b$. Furthermore, as $d\psi$ is taken to be the differential form $\partial_\mu\psi dx^\mu$, the connection A is also a differential form $A_\mu dx^\mu$ or $A^a{}_{b\mu}dx^\mu$.)

In its matrix properties we shall always assume that A is an anti-Hermitian matrix so that $A^\dagger = -A$. This simplifies some of the computations and is no restriction when the gauge group is $O(n)$, $SU(n)$, or any of their subgroups. For the electromagnetic field it means that our A is the quantity usually written $i(e/\hbar c)A_\mu dx^\mu$. Under a gauge transformation corresponding to equations (2.4, 2.5), the transformations of A and its curvature F are

$$\begin{gathered} A \rightarrow g\,A\,g^\dagger - dg\,g^\dagger \,, \\ F = dA + A \wedge A \rightarrow g\,F\,g^\dagger \,. \end{gathered} \tag{2.7}$$

Here the wedge symbol "$\wedge$" indicates the exterior product (i.e., antisymmetrized on the spacetime indices or differential forms, and the d in dA is exterior (antisymmetrized) differentiation.

The harmonic mapping field ϕ in these examples will be a field of matrices $\phi^a{}_b$ that are gauge covariant on both indices. Thus for the ϕ field, the gauge transformation and covariant differentiation rules are

$$\phi \rightarrow g\,\phi\,g^\dagger \tag{2.8}$$

and

$$D\phi = d\phi + A\phi - \phi A = dx^\mu(\nabla_\mu\phi)\,. \tag{2.9}$$

The ϕ field comes to live in a curved manifold when we choose to require that it be a projection matrix satisfying

$$\phi^2 = \phi\,, \qquad \phi = \phi^\dagger\,, \qquad \text{trace}\ \phi = p\,. \tag{2.10}$$

Without these conditions the space of ϕ's could be assigned a flat metric

$$ds'^2 = \frac{1}{2}\ \text{trace}\ [d\phi^\dagger\ d\phi]\,. \tag{2.11}$$

but now equations (2.10) define a curved surface in this high dimensional flat space, and it is the curved geometry of the ϕ surface, obtained by combining equations (2.10) and (2.11), that gives the manifold M' in this example.

The reason for wanting ϕ to be a projection operator is so that any ϕ field will pick out a preferred direction or subspace, thus breaking the original symmetry described by the group G. Given a ϕ field, there exists at any point a preferred subgroup of G, namely the group of g's in G that satisfy $g\ \phi(x)\ g^\dagger = \phi(x)$, so that the existence of a ϕ field gives a preferred role to this reduced gauge symmetry group H. This will appear more clearly in the explicit examples. The detailed form of the action integrals is

$$\begin{aligned} I_A &= -\left(\frac{1}{8e^2}\right) \int d^4x\ \text{trace}\ [F_{\alpha\beta}{}^\dagger\ F^{\alpha\beta}] \\ I_\phi &= -\left(\frac{\mu^2}{4e^2}\right) \int d^4x\ \text{trace}\ [\nabla_\alpha\phi\ \nabla^\alpha\phi]\,. \end{aligned} \tag{2.12}$$

Here $e^2 = e^2/\hbar c$ is a Yang-Mills coupling constant, and μ is a mass.

Abelian SO(2) Example. The simplest example of this class is obtained by letting the gauge group G be $SO(2)$. Although this is merely the Abelian circle group (i.e., rotations in the Euclidean plane about a fixed origin), we shall conform to the notation just introduced and treat its elements g as 2×2 matrices. Then A is an antisymmetric matrix of differential forms

$$A = \begin{pmatrix} 0 & 1 \\ -1 & 0 \end{pmatrix} (A_\beta dx^\beta) \tag{2.13}$$

that is determined by a single vector field A_β. A matrix ϕ that projects any isovector ψ^a onto a line (subspace) making an angle θ with the $a = 1$ axis is defined by

$$\phi_{ab} = \phi_{ba} = \begin{pmatrix} \cos^2\theta & \cos\theta\ \sin\theta \\ \cos\theta\ \sin\theta & \sin^2\theta \end{pmatrix}\,. \tag{2.14a}$$

The manifold M' of ϕ's is then one-dimensional with coordinate θ, although it is obtained embedded in the three-dimensional flat space of all symmetric 2 $\times$

2 matrices. The integer p in equation (2.10) is the dimension of the subspace onto which ϕ projects any isovector ψ, and from the form of ϕ given here one readily verifies equations (2.10) with $p = 1$.

Straightforward evaluation of equation (2.9) with the A and ϕ matrices given here then yields the gauge covariant derivative of ϕ in this example, namely

$$D\phi = d\phi + A\phi - \phi A$$

$$= \begin{pmatrix} -\sin 2\theta & \cos 2\theta \\ \cos 2\theta & \sin 2\theta \end{pmatrix} (d\theta - A_\mu dx^\mu) . \tag{2.14b}$$

Upon substituting these values into the action integrals (2.12), one finds

$$\begin{aligned} \mathrm{I} = \mathrm{I}_A + \mathrm{I}_\phi = -\left(\frac{1}{4e^2}\right) \int d^4x [(\partial_\alpha A_\beta - \partial_\beta A_\alpha)^2 \\ + 2\mu^2(\partial_\beta\theta - A_\beta)^2] . \end{aligned} \tag{2.15}$$

Although $D\phi$ contained many factors nonlinear in θ, these all have disappeared through trigonometric identities when the trace sums in equations (2.12) were carried out.

Thus, this special case with an Abelian group leads to a simple quadratic Lagrangian for a linear theory. It remains an interesting example, however, because a Higgs phenomenon (cf. Bernstein, 1974) occurs—the broken symmetry promotes the originally massless Yang-Mills vector field to a massive vector. To see this we need merely define

$$eC_\beta = \partial_\beta\theta - A_\beta \tag{2.16a}$$

to see that the action integral for the coupled A and ϕ fields takes the form

$$\mathrm{I} = -\left(\frac{1}{4}\right) \int d^4x \, [(\partial_\alpha C_\beta - \partial_\beta C_\alpha)^2 + 2\mu^2 C_\alpha{}^2] . \tag{2.16b}$$

In this simplest example of symmetry breaking using harmonic projection fields, no reduced gauge group remains since $O(1)$ is discrete, and there are no nonlinearities because the original gauge group was Abelian.

Non-Abelian SU(2) *Example.* For the next example we take $G = SU(2)$ as the gauge group and break this symmetry in such a way that a reduced gauge group $H = U(1)$ is the unbroken reduced symmetry. In this case we again see a Higgs phenomenon occur, but one of the original $SU(2)$ gauge vector fields does survive as a massless "photon" field. The other two components become charged massive vector mesons that have nonlinear interactions. The action integral is again given by equations (2.12), but the fields in this case are matrices with complex entries. The projection matrix ϕ is chosen, again with $p = 1$, to project onto a subspace containing an isovector ξ by writing

$$\phi = |\xi> |\xi|^{-2} < \xi| \quad \text{or} \quad \phi^a{}_b = \xi^a (\xi^c \xi^*_c)^{-1} \xi^*_b . \tag{2.17}$$

This preferred direction ξ can be parameterized by two angles $\alpha\beta$ used to rotate a basis vector into an arbitrary direction:

$$\xi = \begin{pmatrix} e^{i\beta/2} & 0 \\ 0 & e^{-i\beta/2} \end{pmatrix} \times \begin{pmatrix} \cos(\alpha/2) & -\sin(\alpha/2) \\ \sin(\alpha/2) & \cos(\alpha/2) \end{pmatrix} \begin{pmatrix} 1 \\ 0 \end{pmatrix} . \tag{2.18a}$$

Then one finds

$$\phi = \frac{1}{2} \begin{pmatrix} 1 + \cos\alpha & e^{i\beta} \sin\alpha \\ e^{-i\beta} \sin\alpha & 1 - \cos\alpha \end{pmatrix} , \tag{2.18b}$$

from which the metric quadratic form can then be computed:

$$ds'^2 = \frac{1}{2} \text{trace } [d\phi d\phi] = \frac{1}{4} (d\alpha^2 + \sin^2\alpha \, d\beta^2) . \tag{2.19}$$

The form with covariant derivatives $D\phi$ must, of course, be used to form the harmonic mapping action I_ϕ from equations (2.12).

After some computation and a choice of gauge, where the field ϕ reduces to the simple form $\phi^a{}_b = \delta^a{}_1 \delta^1{}_b$, one can obtain a relatively compact form for the action integral. In it the matrix A is written in terms of one complex vector field C and one real vector $a = a_\mu dx^\mu$, by writing

$$iA = \begin{pmatrix} a & eC \\ eC^* & -a \end{pmatrix} . \tag{2.20}$$

Gauge transformations of the form

$$g = \begin{pmatrix} e^{i\lambda} & 0 \\ 0 & e^{-i\lambda} \end{pmatrix} \tag{2.21}$$

preserve the special form for ϕ and constitute the reduced gauge group H in this example; the gauge transformation rules for a and C are

$$a \to a + d\lambda , \qquad C \to e^{2i\lambda} C . \tag{2.22}$$

The harmonic mapping term in the action with these notations becomes just $I_\phi = -(\mu^2/2) \int C_\alpha{}^* C^\alpha d^4x$, because $\alpha = 0$ and $\beta = 0$ are constants by the choice of gauge and the only contribution comes from the covariant derivative terms $A\phi - \phi A$ in $D\phi$.

Now turn to the $I_A = -(1/8e^2) \int \mathrm{tr}(F^\dagger \wedge {}^*F)$ part of the action. The curvature $F = dA + A \wedge A$ may be computed from equation (2.20), and one finds that the off-diagonal element F_{12} is given by

$$\left(\frac{i}{e}\right) F_{12} = dC - 2ia \wedge C \equiv D'C .$$

This shows a new (reduced) covariant derivative

$$\nabla'_\mu C_\nu = \partial_\mu C_\nu - 2ia_\mu C_\nu \tag{2.23}$$

that is gauge invariant only under the restricted gauge transformations (2.21) and has a connection specified by a_μ rather than the full matrix A_μ. These terms in the action then give, in $|D'C|^2$, the kinetic energy of a doubly charged vector meson for which the harmonic mapping action supplied the mass term. The diagonal elements of F are

$$iF_{11} = -iF_{22} = da + ie^2 C^* \wedge C$$

and contain terms that give not only the curvature $F' = da$ of the reduced connection, but also terms that give nonminimal couplings between C_μ and a_μ in the action $I_A + I_\phi$.

Combining the parts treated above, then, one finds that in these restricted gauges where $\alpha = 0 = \beta$, the action can be written

$$I = I_A + I_\phi = \int \pounds \, d^4x$$

with

$$\begin{aligned} \pounds = &-\left(\frac{1}{2e^2}\right) [(\partial_\alpha a_\beta - \partial_\beta a_\alpha)\, \partial^\alpha a^\beta] \\ &-\left(\frac{1}{2}\right) [\nabla'_\alpha C_\beta{}^* (\nabla'^\alpha C^\beta - \nabla'^\beta C^\alpha) + \mu^2 C_\alpha{}^* C^\alpha \\ &+ 2i\, (\partial_\alpha a_\beta - \partial_\beta a_\alpha)\, C^{\alpha *} C^\beta] \\ &-\left(\frac{e^2}{4}\right) (C_\alpha{}^* C^\alpha)^2 \end{aligned} \tag{2.24}$$

or, more schematically,

$$\pounds = -\left(\frac{1}{4e^2}\right) F'^2$$

$$-\left(\frac{1}{4}\right) \mid D'C \mid^2 - \left(\frac{\mu^2}{2}\right) \mid C^2 \mid \; - \; iF'_{\alpha\beta} \, C^{\alpha *} \, C^{\beta} \qquad (2.25)$$
$$-\left(\frac{e^2}{2}\right) \mid C \mid^4 .$$

To survey this Lagrangian it is convenient to consider the weak coupling limit, $e^2 \to 0$. Then the e^{-2} term is dominant and, considered alone, requires a_μ to satisfy the source-free Maxwell equations of a massless photon field. The next terms of order e^0, taken alone, describe a massive vector meson field C_α with a double charge (see the equations of 2.22 or 2.23). In addition to the minimal coupling of the C_α field to the background fixed photon field a_μ in the covariant derivative ∇', there is an additional, nonminimal coupling term $-iF'_{\alpha\beta}C^{\alpha *}C^{\beta}$ that is required to make the meson field equation consistent with the divergence condition $\nabla'_\alpha C^\alpha = 0$ when $F \neq 0$ but $\partial_\beta F^{\beta\alpha} = 0$. The high order term $e^2 \mid C \mid^4$ is required to sustain the $\nabla'_\alpha C^\alpha = 0$ condition when the photon field is no longer source free, so $\partial_\beta F^{\beta\alpha} \neq 0$, but has the C^α current as its source. Thus the harmonic projection field ϕ, although gauged to a fixed form $\phi^a{}_b = \delta_1{}^a\delta^1{}_b$, has provided a scheme for generating a vector meson equation with an internal consistency much superior to that provided by minimal coupling ideas.

The action principle in its reduced form (2.24) makes manifest only the gauge symmetries of the equations of (2.22) corresponding to the reduced group $H = U(1)$. However, the field equations remain equivalent to those that would directly follow from the action given in the equations of (2.12) that manifest the larger $G = SU(2)$ group symmetries. These wider symmetries now show themselves mainly in the $\nabla'_\alpha C^\alpha = 0$ harmonic mapping equation. This divergence condition also resembles a Bianchi identity (for the C^α field equations) arising from the no-longer-manifest symmetries described by G/H.

A Pure Harmonic Projection Example. Although the preceding example involved, for its symmetry-breaking field, a field of projection operators or matrices ϕ satisfying a harmonic mapping equation, this harmonic map equation got lost from sight when the special gauge condition on ϕ was imposed. The present example omits the Yang-Mills field and looks explicitly at the harmonic mapping equation for the ϕ field that follows upon varying only the action I_ϕ from equation (2.12) subject to the constraints of (2.10). The manifold of ϕ's can be visualized somewhat by beginning with the Euclidean space of all Hermitian $n \times n$ matrices ϕ endowed with the metric of equation (2.11) or the equivalent vector space norm $\| \phi \|^2 = (1/2)$ trace $\phi^\dagger \phi =$ $(1/2)$ trace ϕ^2. In this vector space, the projection matrices of rank p lie on a surface satisfying $\phi^2 = \phi$ in the hyperplane: trace $\phi = p$. One can easily see that this surface is compact, because it lies on a sphere

$$\| \phi \|^2 = \left(\frac{1}{2}\right) \text{trace } \phi = \frac{p}{2} = \text{const}\,.$$

In carrying out the variation of $I_\phi = -(1/4) \int \text{trace } [\partial_\mu \phi \; \partial^\mu \phi] \; d^4x$, one can make use of the equation

$$d\phi = \phi \; d\phi \; (1 - \phi) + (1 - \phi)\,' d\phi \; \phi \tag{2.26}$$

that holds on any vector tangent to the surface of projection matrices. It is obtained by differentiating the condition $\phi^2 = \phi$ to find $d\phi = \phi d\phi + (d\phi)\phi$ from which $\phi(d\phi)\phi = 0$. Using equation (2.26) then one finds upon varying I_ϕ that the harmonic mapping field equation for a projection field ϕ is

$$(1 - \phi) \; \Delta\phi \; \phi + \phi\Delta\phi \; (1 - \phi) = 0\,,$$

or equivalently (multiply by ϕ)

$$\phi\Delta\phi \; (1 - \phi) = 0\,, \tag{2.27}$$

where $\Delta = -\partial_\alpha \partial^\alpha$.

The field equation (2.27) for harmonic projection fields is in a form that shows similarities to the Einstein equations as mentioned at equations (1.3). Yet it is an equation that formed part of the system studied (by changing field names and gauges) in the previous example that suggested aspects of standard particle physics. It is such hints of unexplored closer relationships in structure and style between gravitation and particle physics that, to me, make a further study of harmonic maps very interesting.

3. SOME OPEN QUESTIONS

Listed here, briefly, are several lines of study that appear interesting. First, can a well-defined quantum theory be achieved for field equations that include (nonlinear) harmonic mappings equations? I see two approaches to this question, and both presume that a straightforward perturbation expansion using the action (0.1) would be nonrenormalizable. (For references on renormalization, see Bernstein, 1974; Bjorken and Drell, 1965; Taylor, 1976.) One approach, applicable to the case of harmonic projection fields as in the preceding examples, would study harmonic maps as limits of renormalizable "relaxed" theories. These theories are obtained by removing the nonlinear constraits $\phi^2 = \phi$ of the equations of (2.10) and instead adding a Higgs polynomial potential

$$-V = -(\mu^2 \, m_H{}^2/8e^2) \text{ trace } [(\phi^2 - \phi)^2]$$

to the action. If the $m_H{}^2 \to \infty$ limit exists for S-matrix elements in this theory, they could be taken to define the harmonic mapping quantum theory. An-

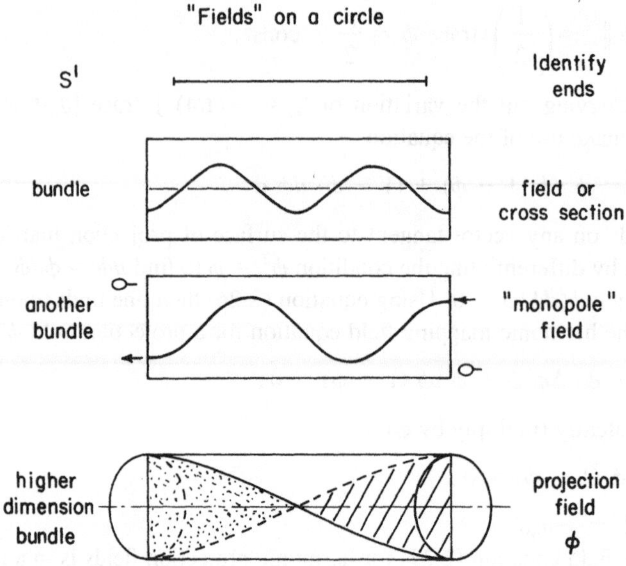

Figure 4.3. A bundle may be constructed in different ways over a circle S^1 as a base manifold. One way of identifying ends gives a simple cylinder; another way gives a Möbius band, and fields like the resulting "monopole" field cannot be smoothly taken to zero. In a higher dimensional situation, a harmonic projection field may pick out a twisted subbundle, whose structure may be different for different solutions ϕ in the same initial bundle.

other approach would be to attempt a rigorous construction of Feynman functional integrals. The chance that this might succeed is of course slim. But the harmonic maps contain some examples much simpler than general relativity, and the Lagrangian is simply a quadratic form in the derivatives, so a serious search for already available mathematical tools to assist with this problem might produce some encouragement.

One should also investigate the possibility that the example cast into the form of equation (2.24) may be renormalizable as it stands. This Lagrangian has merely quartic interactions. The question is whether the divergence condition $\nabla_\alpha C^\alpha = 0$ can be used to control the terms that cause trouble for more general interactions of a massive vector field.

A second set of questions has to do with the relevance of harmonic maps to particle physics. As the examples suggested, harmonic maps can naturally give rise to symmetry breaking. The question is whether the symmetry breaking that is desired on phenomenological grounds can be expressed simply or naturally in terms of harmonic maps or relaxed (Higgs) theories oriented toward a limiting harmonic mapping theory.

A third area of study could be called "dynamical structuring of bundles." The problem to be attacked here is the topological character of magnetic monopoles and their analogues in other bundles. If the electromagnetic field A_μ is seen as a connection in a $U(1)$-bundle then, for a bundle of fixed topological bundle structure, all solutions of Maxwell's equations assigning connections in this bundle have the same monopole charges. One might prefer in some applications to see different values of the monopole analogue arise as different excitation states of a single field, rather than as a priori geometrical constants built into the bundle on which the field equations are imposed. Harmonic projections offer a way to do this.

Consider a solid torus $S^1 \times R^2$ as a bundle of discs ($D^2 \approx R^2$) over a circle S^1. A harmonic projection field ϕ could then pick out preferred R^1 subspaces on each R^2 fiber, and thus define a subbundle over S^1. For some solutions of the harmonic mapping equation this subbundle will be a simple cylinder $S^1 \times R^1$, but for others it will be a Möbius band, depending on how many half twists the preferred axis R^1 in R^2 describes as one scans around the S^1 base space (see Figure 4.3). In a similar way, any harmonic projection field ϕ defines a subbundle whose topological structure may be different for different solutions ϕ in the same initial bundle. In a theory with harmonically broken symmetries then, the Feynman functional integration would automatically include contributions from all the different topological structures that could arise in the reduced bundles.

ACKNOWLEDGMENT

This work was supported in part by National Science Foundation Grants PHY 78-09658 and PHY 79-09281.

REFERENCES

Almgren, F., *Plateau's Problem: An Invitation to Varifold Geometry* (W. A. Benjamin, Inc., New York, 1966).

Appelquist, T., and Bernard, C., "Strongly Interacting Higgs Bosons," *Phys. Rev. D* **22**, 200–213 (1980).

———, "Nonlinear σ Model in the Loop Expansion," *Phys. Rev. D* **23**, 425–438 (1981).

Bernstein, J., "Spontaneous Symmetry Breaking, Gauge Theories, the Higgs Mechanism and All That," *Rev. Mod. Phys.* **46**, 7–48 (1974).

Bjorken, J. D., and Drell, S. D., *Relativistic Quantum Fields* (McGraw-Hill Book Co., New York, 1965).

Coleman, S., " Secret Symmetry: An Introduction to Spontaneous Symmetry Breakdown and Gauge Fields," in *Laws of Hadronic Matter*, part A (edited by A. Zichichi, Academic Press, New York, 1975), pp. 138–223.

DeWitt, B. S., *Dynamical Theory of Groups and Fields* (Gordon and Breach, New York, 1965).

———, "Quantum Theory of Gravity. II. The Manifestly Covariant Theory," *Phys. Rev.* **162**, 1195–1239 (1967).

DeWitt, B. S., Estabrook, F., Wahlquist, H., Christensen, S., Smarr, L., Tsiang, E., "Maximally Slicing a Black Hole," *Phys. Rev. D* **7**, 2814–2817 (1973).

Duff, M. J., and Isham, C. J., "Form-factor Interpretation of Kink Solutions to the Nonlinear σ Model," *Phys. Rev. D* **16**, 3047–3059 (1977).

Eells, J., and Lemaire, L., "A Report on Harmonic Maps," *Bull. Lond. Math. Soc.* **10**, 1–68 (1978).

Eells, J., and Sampson, J. H., "Harmonic Mappings of Riemannian Manifolds," *Am. J. Math.* **86**, 109–160 (1964).

Faddeev, L., and Popov, V. N., "Feynman Diagrams for the Yang-Mills Field," *Phys. Lett. B* **25**, 29–30 (1967).

Faddeev, L., and Slavnov, A. A., *Gauge Fields: Introduction to Quantum Theory* (Benjamin/Cummings Pub. Co., Inc., Reading, Massachusetts, 1980).

Feynman, R. P., "Problems in Quantizing the Gravitational Field, and the Massless Yang-Mills Field," in *Magic Without Magic* (edited by J. Klauder, Freeman and Company, San Francisco, 1972), pp. 377–408.

Fuller, F. B., "Harmonic Mappings," *Proc. Natl. Acad. Sci. USA* **40**, 987–991 (1954).

Gell-Mann, M., and Lévy, M., "The Axial Vector Current in Beta Decay," *Nuovo Cimento (X)* **16**, 705–726 (1960).

Helgason, S., *Differential Geometry, Lie Groups, and Symmetric Spaces* (Academic Press, New York, 1978).

Jackiw, R., "Quantum Meaning of Classical Field Theory," *Rev. Mod. Phys.* **49**, 681–706 (1977).

Mandelstam, S., "Feynman Rules for Electromagnetic and Yang-Mills Fields from the Gauge-independent Field-theoretic Formalism," *Phys. Rev.* **175**, 1580–1603 (1968).

Matzner, R. A., and Misner, C. W., "Gravitational Field Equations for Sources with Axial Symmetry and Angular Momentum," *Phys. Rev.* **154**, 1229–1232 (1967).

Misner, C. W., "Harmonic Maps as Models for Physical Theories," *Phys. Rev. D* **18**, 4510–4524 (1978).

Misner, C. W., Thorne, K. S., and Wheeler, J. A., *Gravitation* (Freeman and Co., San Francisco, 1973).

Moriyasu, K., "Gauge Invariance Rediscovered," *Am. J. Phys.* **46**, 274–278 (1978).

———, "Breaking of Gauge Symmetry: A Geometrical View," *Am. J. Phys.* **48**, 200–204 (1980).

Nutku, Y., "Geometry of Dynamics in General Relativity," *Ann. Inst. Henri Poincaré Sect. A* **21**, 175–183 (1974).

Nutku, Y., and Halil, M., "Colliding Impulsive Gravitational Waves," *Phys. Rev. Lett.* **39**, 1379–1382 (1977).

Penrose, R., and MacCallum, M. A. H., "Twistor Theory: An Approach to the Quantisation of Fields and Space-time," *Phys. Rep.* **6C**, 241–315 (1973).

Scott, A. C., "A Nonlinear Klein-Gordon Equation," *Am. J. Phys.* **37**, 52–61 (1969).

Steenrod, N., *The Topology of Fibre Bundles* (Princeton Univ. Press, Princeton, N.J., 1951).

Taylor, J. C., *Gauge Theories of Weak Interactions* (Cambridge Univ. Press, Cambridge, 1976).

Teitelboim, C., "The Hamiltonian Structure of Spacetime," in *General Relativity and Gravitation* (edited by A. Held, Plenum Press, New York, 1980), pp. 195–225.

Yang, C. N., and Mills, R. L., "Conservation of Isotopic Spin and Isotopic Gauge Invariance," *Phys. Rev.* **96**, 191–195 (1954).

5. Gravitational Fields in General Relativity

ROY P. KERR

1. INTRODUCTION

In the past twenty years there have been many exact solutions found for the field equations of general relativity, but these have proved fairly useless except when it has been possible to investigate their topology, geodesics, asymptotic behavior, and other such properties.

Much effort has gone into investigating the cosmological solutions of the field equations for various physical problems, such as universes containing dust, electromagnetism, or even a Dirac field. When the space is four-dimensionally homogeneous, group theory can be used to reduce the field equations to algebraic ones, usually no worse than quadratic in the unknowns. Unfortunately, these solutions cannot approximate the actual universe, since they are stationary in time as well as being homogeneous in space, and so there is no evolution possible.

By far the most important problem of the last decade has been the investigation of spatially homogeneous spaces evolving in time. For these spaces the field equations simplify to ordinary differential equations, but these are usually so complicated that few exact solutions are known. Nevertheless, the general behavior of the solutions of these equations has been studied to see how the spaces evolve both initially and in the far future. The most important question is whether they have to become singular, and if so whether they collapse in one, two, or three dimensions.

The second most interesting class of solutions are those for isolated gravitating sources, the best known of which is that of Schwarzschild, containing a spherically symmetric body. The theorem of Birkhoff states that there can be no evolution in time for spherically symmetric metrics, and so they have four-dimensional symmetry groups.

For each infinitesimal symmetry of a metric, there exists a corresponding Killing vector satisfying $\xi_{(i,j)} = 0$. If $s \to x^i(s)$ is a geodesic then

$$\frac{d}{ds}(\xi_i v^i) = \xi_{i,j} v^i v^j + \xi_i \frac{dv^i}{ds} = 0,$$

so that $\xi_i v^i$ is a first integral of the geodesic equation. Since the symmetry group of the Schwarzschild metric is four-dimensional, there are plenty of first integrals available, and so the geodesics are very well understood for this geometry.

Weyl investigated the more general problem of axially symmetric static sources where there exist coordinates for which the metric is diagonal and independent of two of the coordinates, one timelike and the other corresponding to a rotation about an axis. The field equations for these are quite simple and the solutions are fairly well understood. Unfortunately, they are not of great physical interest, because they do not include any rotating bodies.

Let E be a rotating, axially symmetric stationary Einstein space. The metric can be written as

$$d\tau^2 = g_{\mathfrak{a}\mathfrak{b}} dx^{\mathfrak{a}} dx^{\mathfrak{b}} + 2g_{A\mathfrak{a}} dx^A dx^{\mathfrak{a}} + g_{AB} dx^A dx^B, \tag{1.1}$$

where $\mathfrak{a}, \mathfrak{b}, \ldots = 1, 2; A, B, \ldots = 3, 4;$ and g_{ij} is a function of the $(x^{\mathfrak{a}})$ but not the (x^A), so that the ∂_A are Killing vectors. We shall call such a metric quasi-diagonalizable (q.d. for short) if the cross terms can be eliminated. This can only be done by a transformation of the type

$$x^A = x^{A'} + f^A(x^{\mathfrak{a}}),\ x^{\mathfrak{a}'} = x^{\mathfrak{a}}. \tag{1.2}$$

If this transformation is substituted into $d\tau^2$ the equation $g_{\mathfrak{a}'A'} = 0$ can be solved for df^A,

$$df^A = -h^{AB} g_{B\mathfrak{a}} dx^{\mathfrak{a}},$$

where $h^{AB} g_{BC} = \delta^A_C$; therefore, the metric is q.d. if and only if the right-hand side is a perfect differential, that is,

$$(h^{AB} g_{B[\mathfrak{a}})_{,\mathfrak{b}]} = 0. \tag{1.3}$$

Papapetrou (1966) showed that those equations are almost satisfied for E. From the field equations, there are two constants of integration that are obstructions to equation (1.3). However, he showed that if E is asymptotically flat, so that there is an actual axis of rotation, then the metric is singular along this axis unless these constants of integration are zero. As a result, it can be assumed that equation (1.3) is satisfied and that the metrics are q.d. Furthermore, the coordinates $x^{\mathfrak{a}}$ can be chosen so that $g_{\mathfrak{a}\mathfrak{b}}$ is diagonal, or inasmuch as all two-metrics are conformally flat, even so that

$$g_{\mathfrak{a}\mathfrak{b}} = \phi^2 \delta_{\mathfrak{a}\mathfrak{b}}.$$

This reduction is not a good idea in practice, because for almost all known solutions the simplest coordinates require $g_{11} \neq g_{22}$, $g_{12} = 0$.

The field equations for these q.d. axially symmetric solutions were known fairly early on, but proved fairly intractable. The first physical exact solutions of these equations were found by the present author while investigating the so-called algebraically special spaces. The metrics and field equations for these were simplified as much as possible and only then were any symmetry requirements imposed. Among the resulting solutions was a two-parameter generalization of Schwarzschild corresponding to a rotating axially symmetric body (Kerr, 1963; Debney, 1967; Debney, Kerr, and Schild, 1969).

After this, Kinnersley (1969*a*,*b*) looked for all Type D solutions and found that as well as the class that I found there was another large class containing no previously known solutions. These were hideously complicated expressions containing Jacobian elliptic functions. As a result, not much was done with them until the work of Plebanski and Demianski (1976), and then Weir and Kerr (1977) showed that there existed coordinates in which they simplified down to a q.d. form in which each component of the metric is a rational expression. They are given in equation (4.1) of this paper.

In recent years, Tomimatsu and Sato (1973) have found further generalizations of the Kerr metric, some of which have been simplified by Ernst. These are again rational and were obtained by using the Ernst transformations to generate new axially symmetric solutions from the old known ones. This trick has been generalized by Kinnersley and is a process whereby if the original solution is rational, then so are the new ones. Given these solutions, it is necessary that their topologies be investigated, particularly their behavior along the symmetry axis. This is fairly easy when they are presented in rational coordinates, as was first shown by Misner (1963) for the NUT metric, subsequently by Carter (1968), and by Boyer and Lindquist (1967) for Kerr, but it can be an impossible task in more general coordinates.

As a result, Plebanski and I have been investigating techniques that could give us all rational solutions of these equations, planning then to throw away those that were not asymptotically flat. We were able to generate many new *static* ones, but the same methods did not work for the stationary case. As a result, we propose to use modern high-speed computers to search for solutions of various degrees. Of course, this requires setting up the problem so that the right variables, both dependent and independent, are used.

In the remainder of this paper I will give a quick presentation of the analysis that led to the Kerr solution. This will be done by a series of problems that the student should go through so that he can learn the techniques used in this area.

2. NOTATION

The standard spinorial notation and methods will be used, because they have proved useful for algebraically special spaces, but everything done this way can be equally well done by using self-dual bivectors. Let $\{\mathbf{e}_{\alpha\dot{\alpha}}\}$ be a spin (Hermitian) basis for the vector fields on the space E $(i, j, k, \ldots = 1, 2, 3, 4; \alpha, \beta, \ldots = 0, 1)$,

$$\mathbf{A} = \mathrm{A}^i \partial_i = \mathrm{A}^{\alpha\dot{\alpha}} \mathbf{e}_{\alpha\dot{\alpha}} . \tag{2.1}$$

Here we distinguish dotted and undotted indices so that there is no relationship between α and $\dot{\alpha}$. It is the *pair* of indices that labels the components of $\mathbf{A}$. For any spin tensor we write

$$\overline{\mathrm{A}^{\dot{\alpha}\ldots\beta\ldots}} = \overline{\mathrm{A}}^{\alpha\ldots\dot{\beta}\ldots} .$$

A bar denotes complex conjugation, and so $\bar{\mathbf{e}}_{\dot{\alpha}\beta} = \mathbf{e}_{\beta\dot{\alpha}}$, because the base is Hermitian. The vector $\mathbf{A}$ is real if and only if $\overline{\mathrm{A}}^{\dot{\alpha}\beta} = \mathrm{A}^{\beta\dot{\alpha}}$.

The basis is chosen so that the components of the metric are

$$g_{\alpha\dot{\alpha}\beta\dot{\beta}} = g(\mathbf{e}_{\alpha\dot{\alpha}}, \mathbf{e}_{\beta\dot{\beta}}) = \varepsilon_{\alpha\beta}\varepsilon_{\dot{\alpha}\dot{\beta}} , \tag{2.2}$$

where $\varepsilon_{\alpha\beta}$ is the usual Levi-Civita symbol in two dimensions. The basis is null, with $\mathbf{e}_{0\dot{0}}$ and $\mathbf{e}_{1\dot{1}}$ being real and the other two being complex conjugates of each other. Taking the signature of g as -2 (as it must be from equation (2.2)), if $(\mathbf{e}_i)$ is a standard Minkowski basis with $\mathbf{e}_4$ timelike, a suitable choice for $\{\mathbf{e}_{\alpha\dot{\alpha}}\}$ is

$$(\mathbf{e}_{\alpha\dot{\alpha}}) = \frac{1}{\sqrt{2}} \begin{Bmatrix} \mathbf{e}_4 + \mathbf{e}_1 & \mathbf{e}_2 + i\mathbf{e}_3 \\ \mathbf{e}_2 - i\mathbf{e}_3 & \mathbf{e}_4 - \mathbf{e}_1 \end{Bmatrix} . \tag{2.3}$$

This choice also makes both $\mathbf{e}_{0\dot{0}}$ and $\mathbf{e}_{1\dot{1}}$ future pointing and fixes the orientation.

The dual basis for differential one-forms on E is

$$\omega^{\alpha\dot{\alpha}} = \omega^{\alpha\dot{\alpha}}_{\ \ i} dx^i ; \quad \omega^{\alpha\dot{\alpha}}(\mathbf{e}_{\beta\dot{\beta}}) = \delta^{\alpha\dot{\alpha}}_{\beta\dot{\beta}} = \delta^{\alpha}_{\beta}\, \delta^{\dot{\alpha}}_{\dot{\beta}} .$$

Spin indices are raised and lowered with $\varepsilon_{\alpha\beta}$,

$$\zeta^\alpha = \varepsilon^{\alpha\beta}\zeta_\beta , \quad \zeta_\alpha = \zeta^\beta \varepsilon_{\beta\alpha} ,$$

agreeing with the usual convention for ordinary tensors, $A_{\alpha\dot{\alpha}} = g_{\alpha\dot{\alpha}\ \beta\dot{\beta}} A^{\beta\dot{\beta}}$.

If $\mathbf{A}$ is any nonzero vector, then

$$|\,\mathbf{A}\,|^2 = \varepsilon_{\alpha\beta}\varepsilon_{\dot{\alpha}\dot{\beta}}\mathrm{A}^{\alpha\dot{\alpha}}\mathrm{A}^{\beta\dot{\beta}} = \det(\mathrm{A}^{\alpha\dot{\alpha}}) ,$$

and so it is null if and only if its matrix has rank one. This means that

$$A^{\alpha\dot{\alpha}} = \zeta^{\alpha}\overline{\eta}^{\dot{\alpha}} . \tag{2.4}$$

Problem 1: Prove that if **A** is real and null, then η is proportional to ζ and that they can be normalized so that

$$A^{\alpha\dot{\alpha}} = \pm\, \zeta^{\alpha}\overline{\zeta}^{\dot{\alpha}} . \tag{2.5}$$

Also, prove **A** is future pointing if and only if the above sign is positive.

Problem 2: Prove that $\zeta^{\alpha}\overline{\eta}^{\dot{\alpha}}$ and $\rho^{\alpha}\overline{\sigma}^{\dot{\alpha}}$ are orthogonal if and only if they have a common factor.

Problem 3: If $\{\mathbf{e}_{\alpha'\dot{\alpha}'}\}$ is another spin basis that has the same orientation as the original and for which the two real vectors are future pointing, then there exists a complex unimodular matrix S such that

$$\omega^{\alpha'\dot{\alpha}'} = S_{\alpha}^{\alpha'} S_{\dot{\alpha}}^{\dot{\alpha}'}\, \omega^{\alpha\dot{\alpha}} . \tag{2.6}$$

I do not know of any way to prove this except by grinding it out using *problem 2* and the orthogonality relations in equation (2.2).

The orthochronous Lorentz transformations are generated by the spin transformations,

$$\zeta^{\alpha'} = S_{\alpha}^{\alpha'}\zeta^{\alpha}, \quad \det (S_{\alpha}^{\alpha'}) = +1 . \tag{2.7}$$

Therefore, the spin metric is invariant under S. The inverse transformation is

$$S_{\alpha'}^{\alpha} = -\, \varepsilon^{\alpha\beta} S_{\beta}^{\beta'}\, \varepsilon_{\beta'\alpha'}$$

Let Γ be the connection, and $\Gamma_j{}^i = \Gamma^i_{jk} dx^k$ the connection one-forms. Because $dg_{\alpha\dot{\alpha}\beta\dot{\beta}} = 0$, $\Gamma_{\alpha\dot{\alpha}\beta\dot{\beta}} = -\Gamma_{\beta\dot{\beta}\alpha\dot{\alpha}}$.

Problem 4: Prove that if $\bar{\mathbf{A}}$ is a real bivector (antisymmetric second rank tensor) then there exists a unique symmetric spinor, $A_{\alpha\beta} = A_{\beta\alpha}$, such that $A_{\alpha\dot{\alpha}\beta\dot{\beta}} = -A_{\alpha\beta}\varepsilon_{\dot{\alpha}\dot{\beta}} - \bar{\mathbf{A}}_{\dot{\alpha}\dot{\beta}}\varepsilon_{\alpha\beta}$. From this deduce

$$\Gamma_{\beta\dot{\beta}}^{\alpha\dot{\alpha}} = \delta_{\beta}^{\alpha}\overline{\Gamma}_{\dot{\beta}}^{\dot{\alpha}} + \delta_{\dot{\beta}}^{\dot{\alpha}}\Gamma_{\beta}^{\alpha} , \quad \Gamma_{\alpha\beta} = \Gamma_{\beta\alpha} = \varepsilon_{\rho\alpha}\Gamma_{\beta}^{\rho} , \tag{2.8}$$

so that taking the covariant differential of any spin tensor as

$$\nabla T_{\beta\cdots}^{\dot{\alpha}\cdots} = dT_{\beta\cdots}^{\dot{\alpha}\cdots} + \overline{\Gamma}_{\dot{\rho}}^{\dot{\alpha}}\, T_{\beta\cdots}^{\dot{\rho}\cdots} - \Gamma_{\beta}^{\rho}\, T_{\rho\cdots}^{\dot{\alpha}\cdots} + \ldots ,$$

gives the usual result for ordinary tensors.

Problem 5: Using *problem 4*, show that if E is empty ($R_{ij} = 0$) then the Riemann tensor is given by

$$R_{\alpha\dot{\alpha}\beta\dot{\beta}\gamma\dot{\gamma}\delta\dot{\delta}} = \Psi_{\alpha\beta\gamma\delta}\varepsilon_{\dot{\alpha}\dot{\beta}}\varepsilon_{\dot{\gamma}\dot{\delta}} + \overline{\Psi}_{\dot{\alpha}\dot{\beta}\dot{\gamma}\dot{\delta}}\varepsilon_{\alpha\beta}\varepsilon_{\gamma\delta} , \tag{2.9}$$

where Ψ is completely symmetric. The standard notation for the five complex components of Ψ is

$$\Psi_0 = \Psi_{0000}, \quad \Psi_1 = \Psi_{0001}, \quad \ldots, \quad \Psi_4 = \Psi_{1111}. \tag{2.10}$$

If $S^{\ \alpha}_{\alpha'}$ is a spin transformation, then $\Psi_{0'} = \Psi_{\alpha\beta\gamma\delta} S^{\ \alpha}_{0'} S^{\ \beta}_{0'} S^{\ \gamma}_{0'} S^{\ \delta}_{0'}$, and so it is zero if and only if the corresponding basis spinor, $S^{\ \alpha}_{0'}$, satisfies a quartic equation. To each root of this equation there corresponds a null vector $\mathbf{e}_{0'\dot{0}'}$. These are the principal null vectors (p.n.v.'s) of Debever-Penrose. If the quartic equation $\Psi_{0'} = 0$ has repeated roots, so that $\Psi_{1'} = 0$ (prove this), then the space is called "algebraically special." It will be assumed hereafter that E is such a space, and that $\mathbf{e}_{0\dot{0}}$ is such a p.n.v.

The remaining spin transformations can be parameterized by two complex functions, $\mathfrak{a}$ and $\mathfrak{s}$,

$$(S^{\alpha'}_{\alpha}) = \begin{Bmatrix} \mathfrak{s}^{-1} & \mathfrak{a}\mathfrak{s}^{-1} \\ 0 & \mathfrak{s} \end{Bmatrix} \tag{2.11}$$

and so, from equation (2.6),

$$\begin{aligned} \omega^{1'\dot{1}'} &= \mathfrak{s}\bar{\mathfrak{s}}\,\omega^{1\dot{1}} \\ \omega^{0'\dot{1}'} &= (\bar{\mathfrak{s}}/\mathfrak{s})\,(\omega^{0\dot{1}} + \mathfrak{a}\omega^{1\dot{1}}) \\ \omega^{0'\dot{0}'} &= (\mathfrak{s}\bar{\mathfrak{s}})^{-1}(\omega^{0\dot{0}} + \mathfrak{a}\omega^{1\dot{0}} + \bar{\mathfrak{a}}\omega^{0\dot{1}} + \mathfrak{a}\bar{\mathfrak{a}}\omega^{1\dot{1}}). \end{aligned} \tag{2.12a}$$

If $\mathbf{e}_{i'} = A^{i}_{i'}\mathbf{e}_i$ is any change of base, the transformation laws for the connection forms are

$$\Gamma^{i'}_{j'} = A_i^{\ i'} A^{\ j}_{j'} \Gamma_j^{\ i} - dA_k^{\ i'} A^{k}_{j'},$$

from which it is easily shown that

$$\Gamma^{\alpha'}_{\beta'} = S^{\alpha'}_{\alpha} S^{\beta}_{\beta'} \Gamma^{\alpha}_{\beta} - dS^{\alpha'}_{\rho} S^{\rho}_{\beta'}.$$

From equation (2.11),

$$\begin{aligned} \Gamma_{0'0'} &= \mathfrak{s}^2\Gamma_{00}, \\ \Gamma_{0'1'} &= \Gamma_{01} - \mathfrak{a}\Gamma_{00} + d(ln\mathfrak{s}), \\ \Gamma_{1'1'} &= \mathfrak{s}^{-2}(\Gamma_{11} - 2\mathfrak{a}\Gamma_{01} + \mathfrak{a}^2\Gamma_{00} - d\mathfrak{a}). \end{aligned} \tag{2.12b}$$

Because $\omega_{1\dot{1}}$ is a p.n.v., $\Psi_0 = \Psi_1 = 0$, and the remaining conformal tensor components transform as

$$\begin{aligned} \Psi_{2'} &= \Psi_2 \\ \Psi_{3'} &= (\Psi_3 - 3\mathfrak{a}\Psi_2)\mathfrak{s}^{-2} \\ \Psi_{4'} &= (\Psi_4 - 4\mathfrak{a}\Psi_3 + 6\mathfrak{a}^2\Psi_2)\mathfrak{s}^{-4}. \end{aligned} \tag{2.12c}$$

These spaces are classified according to the degeneracy of the p.n.v.:

$$\begin{aligned}&\text{Type II: } \Psi_2 \neq 0\\&\text{Type III: } \Psi_2 = 0,\ \Psi_3 \neq 0\\&\text{Type N: } \Psi_2 = \Psi_3 = 0,\ \Psi_4 \neq 0\\&\text{Type D: } 3\Psi_4\Psi_2 - \Psi_3^2 = 0,\ \Psi_2 \neq 0\,.\end{aligned} \tag{2.13}$$

The last type corresponds to a second double p.n.v. being present.

Problem 6: Show that the Cartan equations reduce to

$$d\omega^{\alpha\dot{\alpha}} - (S^{\dot{\alpha}}_{\dot{\beta}}\Gamma^{\alpha}_{\beta\gamma\dot{\gamma}} + \delta^{\alpha}_{\beta}\overline{\Gamma}^{\dot{\alpha}}_{\dot{\beta}\gamma\dot{\gamma}})\omega^{\beta\dot{\beta}} \wedge \omega^{\gamma\dot{\gamma}} = 0 \tag{2.14a}$$

$$d\Gamma_{\alpha\beta} + \Gamma_{\alpha\rho}\wedge\Gamma^{\rho}_{\beta} = \frac{1}{2}\Psi_{\alpha\beta\gamma\delta}\,\omega^{\gamma}{}_{\dot{\rho}}\wedge\omega^{\delta\dot{\rho}} \tag{2.14b}$$

when *E* is empty.

3. COORDINATES AND FIELD EQUATIONS

We shall outline the introduction of standard coordinates and the calculation of the field equations for these spaces.

From equation (2.12*b*) the differential form Γ_{00} is almost an invariant for *E*. Its components are the optical scalars of Sachs for the null vector $\omega^{1\dot{1}}$ (Jordan, Ehlers, and Sachs, 1961),

$$\kappa = \Gamma_{000\dot{0}} = \text{geodesy},$$

$$\rho = \Gamma_{001\dot{0}} = \text{complex divergence},$$

$$\sigma = \Gamma_{000\dot{1}} = \text{shear}.$$

Both ρ and σ are invariant under a null rotation about $\omega^{1\dot{1}}$ ($\mathfrak{a} \neq 0$, $\mathfrak{s} = 1$ in equation (2.11)), provided that $\kappa = 0$; but the fourth component, $\Gamma_{001\dot{1}}$, has no invariant significance if any of the other components are nonzero.

The Goldberg-Sachs theorem states that $\omega^{1\dot{1}}$ is a p.n.v. for an empty Einstein space if and only if it is geodesic and shear-free,

$$\Psi_0 = \Psi_1 \Leftrightarrow \kappa = \sigma = 0\,,$$

and therefore

$$\Gamma_{00} = \rho\omega^{1\dot{0}} + \Gamma_{001\dot{1}}\omega^{1\dot{1}}\,. \tag{3.1}$$

We shall only consider diverging spaces where $\rho \neq 0$, inasmuch as the others are rather trivial.

ASSUMPTION: $\rho \neq 0$. (3.2)

Now, the (00) component of equation (2.14*b*) reduces to

$$d\Gamma_{00} + 2\Gamma_{00}\Lambda\Gamma_{01} = \Phi_2\omega^{1\dot{0}}\Lambda\omega^{1\dot{1}} \tag{3.3}$$

and so, from equation (3.1),

$$d\Gamma_{00}\Lambda\Gamma_{00} = 0 .$$

This is the equation that allows for introduction of special coordinates into these spaces and the reduction of the field equations to comparatively simple form. Its complete solution is $\Gamma_{00} = \phi d\zeta$, where both ϕ and ζ are complex functions. A spin transformation, equation (2.11), with $\mathfrak{s}^2 = \phi_\zeta$, $\mathfrak{a} = \Gamma_{001\dot{1}}/\rho$, transforms Γ_{00} to $d\zeta$ and $\Gamma_{001\dot{1}}$ to zero, so that

$$\Gamma_{00} = \rho\omega^{1\dot{0}} = d\zeta . \tag{3.4}$$

Any transformation preserving this must satisfy $d\zeta' = \mathfrak{s}^2 d\zeta$, so that

$$\zeta' = \Phi(\zeta) , \quad \mathfrak{s}^2 = \Phi_\zeta , \quad \Gamma_{001\dot{1}} = 0 , \tag{3.5}$$

where Φ is an analytic function. If the Cartan equations are expanded as equations on the rotation coefficients, then the Newman-Penrose equations are obtained. One of these says that the $-Re(\rho^{-1})$ is an affine parameter along $\omega^{1\dot{1}}$, and so it will be used as a coordinate called v. The two complex functions ζ and $\bar{\zeta}$ are also fairly natural coordinates, so we only need to find one more. The usual one is called u, and it is any solution of $\mathbf{e}_{0\dot{0}}(u) = 0$, $\mathbf{e}_{1\dot{1}}(u) = 1$. This is equivalent to requiring that $\omega^{1\dot{1}} = du + \Omega d\zeta + \bar{\Omega} d\bar{\zeta}$ for some complex function Ω. With $(u, v, \zeta, \bar{\zeta})$ as coordinates, it is fairly easy to show that

$$\begin{aligned} \omega^{1\dot{1}} &= du + \Omega d\zeta + \bar{\Omega} d\bar{\zeta} \\ \omega^{0\dot{0}} &= dv + \beta d\bar{\zeta} + \bar{\beta} d\bar{\zeta} + U\omega^{1\dot{1}} \\ \omega^{1\dot{0}} &= d\zeta/\rho . \end{aligned} \tag{3.6}$$

If the rotation coefficients are calculated from these and then substituted into the Newman-Penrose equations that involve derivatives with respect to v (i.e., in the $\mathbf{e}_{0\dot{0}}$ direction), it can be shown that Ω is independent of v and that β, U and ρ are functions of this and another complex function μ,

$$\begin{aligned} -\beta &= i(\mathbf{D} - \Omega_u)\Delta + v\Omega_u \\ U &= Re[-\mathbf{D}\bar{\Omega}_u - \mu\rho] \\ \rho &= -(v + i\Delta)^{-1} \\ \Delta &= Im(\mathbf{D}\bar{\Omega}) . \end{aligned} \tag{3.7}$$

The operator **D** is essentially the derivative in the direction of $\mathbf{e}_{1\dot{0}}$,

$$\mathbf{e}_{1\dot{0}} = \rho(\mathbf{D} - \beta\partial_v)\,, \qquad \mathbf{D} = \partial_\zeta - \Omega\partial_u\,. \tag{3.8}$$

Because both Ω and the complex mass $\mu = M_1 + iM_2$ are independent of v, the v dependence of the metric is exhibited explicitly.

The metric is already rational in the coordinate v, certain terms having as denominator $(v^2 + \Delta^2)$. The other coordinates are such that even for the simplest solutions the metric is not rational, and it is always necessary to introduce a coordinate quite different from u before the metric has a reasonable form.

Problem 7: Show that the only nonzero components of Γ and Ψ are

$$\begin{aligned}
\Gamma_{001\dot{0}} &= \rho\,,\\
\Gamma_{001\dot{0}} &= \rho\Omega_u\,,\\
\Gamma_{011\mathrm{i}} &= \frac{1}{2}\mu\rho^2\,,\\
\Gamma_{110\mathrm{i}} &= -\bar{\rho}(\bar{\mathbf{D}}\partial_u\Omega) + \frac{1}{2}\mu(\rho^2 + \rho\bar{\rho})\,,\\
\Gamma_{111\mathrm{i}} &= -\bar{\Omega}_{uu} + \rho H + \frac{1}{2}\rho^2 I + \rho^3\mu J\,,
\end{aligned} \tag{3.9a}$$

$$\begin{aligned}
\Psi_2 &= -\mu\rho^3\,,\\
\Psi_3 &= \rho^2 H + \rho^3 I + 3\rho^4\mu J\,,\\
\Psi_4 &= -\rho(\partial_u\partial_u\mathbf{D}\Omega) - \rho^2(\mathbf{D} - 4\Omega_u)H\\
&\quad - \rho^3\left[\frac{1}{2}(\mathbf{D} - 5\Omega_u)I + 2HJ\right]\\
&\quad + \rho^4[(\mathbf{D} - 6\Omega_u)(\mu J) + 2IJ] - 6\mu J^2\rho^5\,,
\end{aligned} \tag{3.9b}$$

where

$$\begin{aligned}
H &= \bar{\mathbf{D}}\partial_u\mathbf{D}\Omega\,,\\
I &= \mathbf{D}\mu - 3\Omega_u\mu\,,\\
J &= \frac{1}{2}(\mathbf{D}\bar{\mathbf{D}}\bar{\Omega} - \bar{\mathbf{D}}\mathbf{D}\Omega) = i(\mathbf{D} - \Omega_u)\Delta\,.
\end{aligned} \tag{3.10}$$

We shall not really use Ψ_3 and Ψ_4, so their calculation is not needed here. The remaining field equations are

$$\mathrm{I}\colon \partial_u\mu - (\bar{\mathbf{D}} - 2\bar{\Omega}_u)H = 0$$

$$\text{II: } Im(\mu + \overline{\mathbf{D}}\mathbf{D}\overline{\mathbf{D}}\mathbf{D}u) = 0 \tag{3.11}$$

$$\text{III: } \overline{\mathbf{D}}\mu - 3\overline{\Omega}_u\mu = 0$$

The first of these can be rewritten as

$$\partial_u(\mu - \overline{\mathbf{D}\mathbf{D}}\mathbf{D}\Omega) = |\, \partial_u D\Omega \,|^2, \tag{3.12}$$

which shows that it is compatible with II. They are necessary and sufficient for the metric

$$d\tau^2 = 2\omega^{0\dot{0}}\,\omega^{1\dot{1}} - 2\,|\,\omega^{0\dot{1}}\,|^2, \tag{3.13}$$

with the $\omega^{\alpha\dot{\alpha}}$ given by equations (3.6,3.7), to be an algebraically special, empty, Einstein space.

Problem 8: Show that the remaining set of coordinate transformations, $\mathfrak{C}$, is

$$\mathfrak{C}\text{: } \zeta' = \phi(\zeta), \tag{3.14a}$$

$$u' = [u + S(\zeta, \bar{\zeta})]\,|\,\Phi_\zeta\,|, \tag{3.14b}$$

$$v' = v/\,|\,\Phi_\zeta\,|, \tag{3.14c}$$

$$\mu' = \mu/\,|\,\Phi_\zeta\,|^3, \tag{3.14d}$$

$$\Omega' = \left[\Omega - S_\zeta - \frac{1}{2}(\Phi_{\zeta\zeta}/\Phi_\zeta)(u + S)\right]|\,\Phi_\zeta\,|\,/\Phi_\zeta, \tag{3.14e}$$

and that the corresponding spin transformation is

$$(\zeta, \mathfrak{a}) = (\Phi_\zeta, 0)\,. \tag{3.15}$$

From the equations of (2.13) and equation (3.9) the various Petrov types are given by

$$\begin{aligned} &\text{Type II: } \mu \neq 0 \\ &\text{Type III: } \mu = 0,\quad \overline{\mathbf{D}}\partial_u\mathbf{D}\Omega \neq 0 \\ &\text{Type N: } \mu = 0,\quad \overline{\mathbf{D}}\partial_u\mathbf{D}\Omega = 0,\quad \partial_u\partial_u\mathbf{D}\Omega \neq 0 \\ &\text{Flat space: } \mu = 0,\quad \partial_u\mathbf{D}\Omega = f(\zeta) \end{aligned} \tag{3.16}$$

The Type D conditions are much more complicated and have been solved completely by Kinnersley (1969*a*,*b*). Subsequently, Plebanski and Demianski (1976) exhibited these metrics in a simple algebraic form and then Weir and Kerr (1977) showed the connection between these and the spaces of Kinnersley.

It is convenient to introduce a less restrictive coordinate system (r, s, u, v) in which $\mathbf{e}_{0\dot{0}}(s) = 0$, but $\mathbf{e}_{1\dot{1}}(s)$ is arbitrary.

$$u = u(s, \zeta, \bar{\zeta}),\ r = vP,\ P = \frac{\partial u}{\partial s}. \tag{3.17}$$

This is the P used by Weir and Kerr, but it is the inverse of that used by Kerr (1963) and by Robinson and coworkers. It is related to the p of Kerr and Debney (1970) by $P = e^{-p}$.

Problem 9: Show that in these coordinates,

$$(d\tau)^2 = -2(\Sigma/P^2)d\zeta d\bar{\zeta} + 2\omega(dr + \beta d\zeta + \bar{\beta} d\bar{\zeta} + U\omega), \tag{3.18}$$

where

$$\begin{aligned}
&\omega = ds + \Lambda d\zeta + \bar{\Lambda} d\bar{\zeta},\\
&\Sigma = r^2 + \delta^2,\ \delta = \Delta P = Im(\mathbf{D}\bar{\Lambda})P^2,\\
&-\beta = (r - i\delta)\Lambda_s + i\mathbf{D}\delta,\\
&U = -\ r_s/P + Re[P^2\mathbf{D}(\bar{\mathbf{D}}\ln P - \bar{\Lambda}_s) + (m_1 r - m_2\delta)\Sigma^{-1}],\\
&\rho = -\ P/(r + i\delta),\ \mathbf{D} = \partial_\zeta - \Lambda\partial_\zeta.
\end{aligned} \tag{3.19}$$

The functions Λ and m are related to Ω and μ by

$$\begin{aligned}
&\Omega = P\Lambda - u_\zeta = -Du,\\
&m = m_1 + i\,m_2 = \mu P^3.
\end{aligned} \tag{3.20}$$

Obviously, the original metric is obtained from this by substituting $P = 1$ and relabeling the functions and coordinates. The field equations will not be transformed to these coordinates because it is just as easy to use equation (3.12) together with the appropriate definitions of the last paragraph.

It was shown in Kerr and Debney (1970) that for the class of axially symmetric stationary metrics that includes Schwarzschild, s can be chosen so that ∂_s is a Killing vector, and therefore the metric is independent of s.

Problem 10: Show that if the metric is independent of s,

$$\begin{aligned}
&(d\tau)^2 = -\ 2(\Sigma/P^2)\ |\ d\zeta - i(P\delta_{\bar{\zeta}}/\Sigma)\omega\ |^{\ 2} + 2\omega dr + (\pi/\Sigma)\omega^2,\\
&\Sigma = r^2 + \delta^2,\ \pi = 2K_1 r^2 = 2Re(m)r - 2K_2,\\
&K_1 = PP_{\zeta\bar{\zeta}} - P_\zeta P_{\bar{\zeta}},\ K_2 = \mathrm{Q}\mathrm{Q}_{\zeta\bar{\zeta}} - \mathrm{Q}_\zeta \mathrm{Q}_{\bar{\zeta}},\\
&\mathrm{Q} = \delta P = P^3 Im(\bar{\Lambda}_\zeta),
\end{aligned} \tag{3.21}$$

and that the field equations are comparatively simple:

$$\begin{aligned}
&\text{I: } K_{1\zeta\bar{\zeta}} = 0,\\
&\text{II: } Im(m) = P^2\delta_{\zeta\bar{\zeta}} + 2K_1\delta,
\end{aligned} \tag{3.22}$$

III: $m_{\bar{\zeta}} = 0$.

From the last of these, $m = m(\zeta)$, and so the second cannot be integrated unless the right-hand side is a harmonic function, which gives a fourth order equation for δ.

There is only one solution known of I when K_1 is nonconstant and that leads to solutions that are not asymptotically flat. It will be discussed in a survey article by Kerr and Robinson. Since K_1 is the curvature of $d\zeta d\bar{\zeta}/P^2$, the case where K_1 is constant is fairly well understood.

Problem 11: Show that the residual group of coordinate transformations that preserves equations (3.21) and leaves the metric independent of s is $\zeta' = \Phi(\zeta)$, $s' = c_0(s + A)$, $r' = c_0{}^{-1}r$, $m' = c_0{}^{-3}m$, $\Lambda' = c_0\Phi_\zeta{}^{-1}(\Lambda - A_\zeta)$, $P' = c_0{}^{-1} \mid \Phi_\zeta \mid P$, where c_0 is a constant and A is a real function of ζ and $\bar{\zeta}$.

Problem 12: When K_1 is constant show that $(P_{\zeta\zeta}/P)_{\bar{\zeta}} = 0$, and deduce that there exists a ζ coordinate for which $P_{\zeta\zeta} = 0$.

We shall use a ζ coordinate for which

$$P_{\zeta\zeta} = 0\,,$$

$$PP_{\zeta\bar{\zeta}} - P_\zeta P_{\bar{\zeta}} = K_1 = \text{constant}\,,$$

and so $P = a\zeta\bar{\zeta} + b\zeta + \bar{b}\bar{\zeta} + c$, a bilinear function of ζ and $\bar{\zeta}$.

Problem 13: Show that any Φ-transformation that preserves the first of these must satisfy

$$(ln\Phi_\zeta)_{\zeta\zeta} = \frac{1}{2}[(ln\Phi_\zeta)_\zeta]^2\,,$$

and show that the complete solution of this is the bilinear transformation

$$\zeta' = \frac{a\zeta + b}{c\zeta + d}\,, \quad \det\begin{pmatrix} a & b \\ c & d \end{pmatrix} = 1\,.$$

In Kerr and Debney (1970) it was shown that if these metrics admit a second Killing vector commuting with ∂_s, then it can be chosen to be $Re(\partial_\zeta)$. If we write

$$d\zeta = \frac{1}{2}(d\phi + i\,d\theta)\,, \quad \partial_\zeta = \partial_\phi - i\,\partial_\theta\,,$$

then the metrical functions can be taken to be functions of θ alone, so that the metric is independent of both ϕ and s. Since Λ is any solution of $Im(\overline{\Lambda}_\zeta) = QP^{-3}$, it can be chosen to be real. The metric is

$$d\tau^2 = -\frac{1}{2}\Sigma P^{-2}\{(d\theta)^2 + [2d\phi + (2\dot{\delta}P^2/\Sigma)(ds + \Lambda d\phi)]^2\} \\ + [2dr + (\pi/\Sigma)(ds + \Lambda d\phi)]\,(ds + \Lambda d\phi)\,, \tag{3.23}$$

where a dot denotes differentiation with respect to θ. For this to be q.d., the appropriate coordinate transformation must eliminate all cross terms between $(d\theta, dr)$ and $(d\phi, ds)$. There is only one such transformation, and therefore if $(s, \phi) \to (\eta_1, \eta_2)$,

$$\begin{aligned} d\phi + (2\dot{\delta}P^2/\Sigma)(ds + \Lambda d\phi) &= d\eta_2 + (2\dot{\delta}P^2/\Sigma)(d\eta_1 + \Lambda d\eta_2) \\ ds + \Lambda d\phi &= d\eta_1 + \Lambda d\eta_2 - (\Sigma/\pi)dr\,, \end{aligned}$$

that is,

$$\begin{aligned} d\eta_1 &= ds + (\Sigma + 2\Lambda\dot{\delta}P^2)dr/\pi\,, \\ d\eta_2 &= d\phi - (2\dot{\delta}P^2/\pi)dr\,, \\ -\,d\tau^2 &= \frac{\Sigma}{2P^2}\{d\phi^2 + [d\eta_2 + (2\dot{\delta}P^2/\Sigma)(d\eta_1 + \Lambda d\eta_2)]^2\} \\ &\quad + \frac{\Sigma}{\pi}dr^2 - \frac{\pi}{\Sigma}(d\eta_1 + \Lambda d\eta_2)^2\,. \end{aligned} \tag{3.24}$$

These equations are integrable if and only if the coefficients of dr are functions of r, and so the ratio $(r^2 + \delta^2 + 2\Lambda\dot{\delta}/P^2)/\dot{\delta}P^2$ must be independent of θ, as must π. From this K_1, K_2, $\dot{\delta}P^2$ and m are all constants. These conditions can be shown to be precisely those needed for E to be Type D.

It is not obvious that II is satisfied, but this follows from the next problem.

Problem 14: Show that II can be written as

$$P^{-2}\,|\,P^2\delta_\zeta\,|^{\,2} + \delta^2 K_1 - Im(m)\delta + K_2 = 0\,, \tag{3.25}$$

and that the following is an identity

$$[(P^2\delta_\zeta)_\zeta/PQ]_{\bar{\zeta}} \;\equiv\; Q^{-2}K_2 \;-\; P^{-2}K_1\,. \tag{3.26}$$

The last of these shows that if $P^2\delta$ and K_1 are both constant then so is K_2. What is remarkable is that it is not really necessary to solve any differential equations, provided the correct coordinates are used.

Suppose that (θ, r) are replaced by two new coordinates (x, y), where

$$x = \delta(\theta),\ y = r\,.$$

The c_0 scale transformation of *problem 11* can be used to set

$$2P^2\delta = 1\,,$$

because $(2P^2\dot{\delta})' = c_0^3(2P^2\dot{\delta})$. The equations of (3.21) can be solved for Λ,

$$\Lambda = -\delta^2/2P^2\dot{\delta} = -\delta^2 = -x^2$$

because from this, $\dot{\Lambda} = -2\delta\dot{\delta}/2P^2\dot{\delta} = -\delta/P^2$. If we define

$$\begin{aligned} \pi_x &= (2P^2)^{-1} = -2(K_2x^2 - Im(m)x + K_2)\,, \\ \pi_y &= \pi = -2(-K_1y^2 + Re(m)y + K_2)\,, \\ \Sigma &= x^2 + y^2\,, \end{aligned} \tag{3.27}$$

(the subscripts of π_x and π_y just distinguish two distinct polynomials), then the metric (3.24) becomes

$$\begin{aligned} -d\tau^2 = {} & \frac{\Sigma}{\pi_x}dx^2 + \frac{\Sigma}{\pi_y}dy^2 + \frac{\pi_x}{\Sigma}(d\eta_1 + y^2 d\eta_2)^2 \\ & - \frac{\pi_y}{\Sigma}(d\eta_1 - x^2 d\eta^2)^2\,. \end{aligned} \tag{3.28}$$

Problem 15: Prove this, using *problem 14* to replace $(2P^2)^{-1}$ in the metric by the polynomial π_x.

Problem 16: Show that the curvature invariant Ψ_2 is given by

$$\Psi_2 = -m(y + ix)^{-3}\,,$$

and deduce that the curvature tensor is regular for all points other than at $(x, y) = (0, 0)$.

The remaining coordinate transformation is

$$\begin{aligned} (x', y', \eta_1', \eta_2') &= (qx, qy, q^{-1}\eta_1, q^{-3}\eta_2)\,, \\ (K_1', K_2', m') &= (q^2K_1, q^4K_2, q^3m)\,, \end{aligned} \tag{3.29}$$

which preserves the form of equation (3.28) but modifies the parameters, showing that they are not invariants. If $2K_1 \neq 0$, it can be chosen to be ± 1. Since both π_x and π_y have to be positive, $-K_2$ must be also whenever $K_1 > 0$, and $Im(m) = 0$. This leads to the Kerr metric

$$\pi_x = a^2 - x^2\,, \qquad \pi_y = y^2 - 2my + a^2\,. \tag{3.30}$$

$Im(m)$ is the NUT parameter and produces metrics that are not asymptotically flat.

The analysis of Misner (1963) for NUT applies equally well to the class here, so these more general solutions do not represent particles (i.e., are not asymptotically flat).

We have been assuming that δ is nonconstant so that it can be used as a coordinate. However, it is zero for Schwarzschild, so the above metrics cannot include this except as some kind of limit. When $\dot{\delta} = 0$, we find $Q = \delta P$, $K_2 = \delta^2 K_1$, and $Im(m) = 2\delta K$ from equation (3.25). A new coordinate, x, is defined by

$$\frac{dx}{d\theta} = \pi_x \stackrel{\text{def}}{=} \frac{1}{2P^2},$$

and the coordinate u is chosen so that $\Lambda = 2\delta x$. From this definition for π_x and the definition of K_1 in terms of P,

$$\frac{d^2\pi_x}{dx^2} + 4K_1 = 0.$$

Problem 17: Show that if $\dot{\delta} = 0$, then the metric in equation (3.24) can be written as

$$-d\tau^2 = \Sigma\left[\frac{dx^2}{\pi_x} + \frac{dy^2}{\pi_y} + \pi_x d\eta_2^2\right] - \frac{\pi_y}{\Sigma}(d\eta_1 + 2\delta x d\eta_2)^2,$$

where

$$(y, \eta_1, \eta_2) = (r, s, \phi),$$

$$\pi_x = -2K_1 x^2 + 2qx + p > 0,$$

$$\pi_y = 2K(y^2 - \delta^2) - 2m_0 y > 0,$$

$$\Sigma = y^2 + \delta^2,$$

$$m = m_0 + 2iK_1\delta,$$

where p and q are arbitrary constants. This is the $B[+]$ metric of Carter and is also called the generalized NUT metric.

Problem 18: If $K_1 < 0$, prove that p and q can both be transformed to zero by taking $P = (\sqrt{-K_1})i(\zeta - \bar{\zeta})$, but this does not work for $K_1 > 0$. In the latter case the best that can be done is to eliminate either p or q.

Also, the same transformation as equation (3.29) can be used to transform any nonzero $|K_1|$ to 1/2, and so the canonical forms for π_x and π_y are

$$K_1 > 0: \pi_x = 1 - x^2, \quad \pi_y = y^2 - 2my - \delta^2;$$

$$K_1 < 0: \pi_x = x^2, \quad \pi_y = -y^2 - 2my + \delta^2;$$

$$K_1 = 0: \pi_x = 1, \quad \pi_y = -2my.$$

The corresponding ranges of coordinates follow from the requirement that both of these polynomials have to be positive. When $K_1 > 0$, x lies between the roots of π_x, that is, $x \varepsilon (-1, 1)$, and y lies between the upper root of $\pi_y = 0$ and $+\infty$.

The curvature invariant Ψ_2 is

$$\Psi_2 = -m(y + i\delta)^{-3},$$

and so it is nowhere singular if $\delta \neq 0$. When δ is zero we get the usual invariants of Schwarzschild, these being singular at $y = 0$.

To see that the Kerr metric is asymptotically flat we first transform to "spherical polar" coordinates,

$$x = a \cos\theta, \quad y = r, \quad \eta_1 = t + a\phi, \quad \eta_2 = \phi/a,$$

so that

$$-d\tau^2 = (r^2 + a^2)\,[(d\theta)^2 + \sin^2\theta(d\phi)^2] + \frac{\Sigma}{\pi y}(dr)^2$$
$$- \frac{\pi y - a^2\sin^2\theta}{\Sigma} \cdot dt^2$$
$$-2a\left(\frac{\pi y - \Sigma}{\Sigma}\right) \sin^2\theta d\phi dt.$$

Problem 19: Transform this to "rectangular" coordinates by the usual equations connecting (r, θ, ϕ) to (x, y, z), and show that the metric is well behaved along the symmetry axis where $\theta = 0$ or π. Essentially, this is because the terms after the first involve $\sin^2\theta\, d\phi$, which becomes $x\, dy - y\, dx$ in "rectangular" coordinates.

4. CONCLUSIONS

What is surprising about this particular class of algebraically special metrics is that the field equations have been solved without actually writing down a solution of the differential equations. These equations were integrated by using them to define new coordinates. Although the original coordinates used for algebraically special spaces are the most obvious, in practice they are usually hopeless for any particular space. The other class of Type D metrics, that of Kinnersley, was first written down as a very complicated expression involving Jacobian elliptic functions. However, both Plebanski and Demianski (1976) and Wier and Kerr (1977) have shown that coordinates exist in which the metric is both q.d. and rational, namely

$$-d\tau^2 = (x - y)^{-2}\left[\left(\frac{dx^2}{g(x)} - \frac{dy^2}{g(y)}\right)\Sigma\right. \tag{4.1}$$

$$+ \frac{g(x)}{\Sigma}(d\eta_1 + y^2 d\eta_2)^2 + \frac{g(y)}{\Sigma}(d\eta_2 - x^2 d\eta_1)^2 \Bigg] ,$$

where

$$g(t) = c_0(1 - t^4) + c_1 t + c_2 t^2 + c_3 t^3,\ t = x, y ,$$
$$\Sigma = 1 + x^2 y^2 .$$

For this there is only one polynomial g corresponding to the pair, π_x and π_y, but it is quartic rather than quadratic. Unfortunately, this general metric is not asymptotically flat. It has a conical singularity that arises because the magnitude of the derivatives of $g(x)$ at two of its neighboring roots are different. This means that the periods of the angular coordinates near the north and south poles of the polar coordinates cannot be chosen consistently.

In general, a lot of effort has been put into finding the few q.d. solutions that are known. Since all of these are rational in the end, I believe that it would be better to use modern high-speed computers to find all possible asymptotically flat rational solutions of a given degree than to use special methods to find particular solutions such as those given here or the Tomimatsu-Sato ones.

REFERENCES

Boyer, R. H., and Lindquist, R. W., "Maximal Analytic Extension of the Kerr Metric," *J. Math. Phys.* **8**, 265–281 (1967).

Carter, B., "Hamilton-Jacobi and Schrodinger Separable Solutions of Einstein's Equations," *Commun. Math. Phys.* **10**, 280–310 (1968).

Debney, G. C., "Killing Vectors in Empty, Algebraically Special Einstein Manifolds with Diverging Ray Congruences," Ph.D. dissertation, The University of Texas (1967).

Debney, G. C., Kerr, R. P., and Schild, A., "Solutions of the Einstein and Einstein-Maxwell Equations," *J. Math. Phys.* **10**, 1842–1854 (1969).

Ernst, F. J., "New Formulation of the Axially Symmetric Gravitational Field Problem," *Phys. Rev.* **167**, 1175–1178 (1968).

Jordan, P., Ehlers, J., and Sachs, R., "Exact Solutions of the Field Equations of General Relativity, II: Contributions to the Theory of Pure Gravitational Radiation," *Akad. Wiss. Lit. Mainz Abh. Math. Nat. Kl* **1**, 3–60 (1961).

Kerr, R. P., "Gravitational Field of a Spinning Mass as an Example of Algebraically Special Metrics," *Phys. Rev. Lett.* **11**, 237–238 (1963).

Kerr, R. P., and Debney, G. C., "Einstein Spaces with Symmetry Groups," *J. Math. Phys.* **11**, 2807–2817 (1970).

Kinnersley, W., "Type D Gravitational Fields," Ph.D. dissertation, The California Institute of Technology (1969*a*).

———, "Type D Vacuum Metrics," *J. Math. Phys.* **10**, 1195–1203 (1969*b*).

Misner, C. W., "The Flatter Regions of Newman, Unti, and Tamburino's Generalized Schwarzschild Space," *J. Math. Phys.* **4**, 924–937 (1963).

Papapetrou, A., "Champs Gravitationnels stationnaires à symétrie axiale," *Ann. Inst. Henri Poincaré* **4**, 83–105 (1966).

Plebanski, J., and Demianski, M., "Rotating, Charged, and Uniformly Accelerating Mass in General Relativity," *Ann. Phys. (USA)* **98**, 98–127 (1976).

Robinson, I., and Robinson, J. R., "Vacuum Metrics Without Symmetry," *Int. J. Theor. Phys.* **2**, 231–242 (1969).

Robinson, I., Robinson, J. R., and Zund, J. D., "Degenerate Gravitational Fields with Twisting Rays," *J. Math. Mech.* **18**, 881–892 (1969).

Robinson, I., and Trautman, A., "Same Spherical Gravitational Waves in General Relativity," *Proc. R. Soc. London Ser. A* **265**, 463–473 (1962).

Tomimatsu, A., and Sato, H., "New Series of Exact Solutions for Gravitational Fields of Spinning Masses," *Prog. Theor. Phys.* **50**, 95–110 (1973).

Weir, G. J., Ph.D. dissertation, The University of Canterbury (1976).

Weir, G. J., and Kerr, R. P., "Diverging Type-D Metrics," *Proc. R. Soc. London Ser. A* **355**, 31–52 (1977).

6. On the Potential Barriers Surrounding the Schwarzschild Black Hole

SUBRAHMANYAN CHANDRASEKHAR

1. INTRODUCTION

The remarkable simplicity of the general relativistic theory of black holes derives from the fact that under stationary conditions, Einstein's equations allow for them only a single two-parameter family of solutions. This is the Kerr family of solutions. The two parameters of the solution are the mass M and the angular momentum J of the black hole. If we enlarge Einstein's equations to the Einstein-Maxwell equations (i.e., Einstein's equations with a source represented by the energy-momentum tensor of an electromagnetic field) then the Kerr family is enlarged to the Kerr-Newman family with an additional parameter Q_* which specifies the charge of the black hole.

When we restrict ourselves to the case $J = 0$, the solutions become spherically symmetric and reduce to the ones associated with the names of Schwarzschild and of Reissner and Nordström.

We have now a fairly complete understanding of the Schwarzschild, the Reissner-Nordström, and the Kerr solutions and their perturbations. Indeed, many of the features of the Reissner-Nordström and Kerr solutions are already present in the Schwarzschild solution in simpler contexts; and an understanding of the Schwarzschild solution appears as a prerequisite to an understanding of the other solutions. For this reason, the present account will be restricted to the Schwarzschild solution.

One of the best ways of understanding the attributes of a physical system is to find out how it reacts to external perturbations and, in the first instance, to infinitesimal perturbations. In the case of the black holes this is the only method available to us since there is no way in which an external observer can explore the "other side" of the horizon.

The reaction of an object to an infinitesimal perturbation is determined by the enumeration of the so-called normal modes of oscillation. In the case of the black holes, this enumeration reduces to finding how a black hole reacts to incident waves of different sorts. The solution to this latter problem bears on a number of related questions: the stability of the black hole and the determination of the "quasi-normal modes" (which describe the long-term behavior of a black hole that has been perturbed) (i.e., to its ringing). Further, the analysis, besides suggesting relations to domains which one might not have expected, discloses what appears to be a hidden symmetry of the solutions. Since a complete and self-contained account, of all these different aspects, is not possible within the limits of a single article (even when restricted to the Schwarzschild black hole), we must content ourselves to a bare description of the methods and the results except in cases where no ready reference is available.

2. AN OUTLINE OF THE PROBLEM

It is known that by choosing a coordinate frame consistent with the assumption of spherical symmetry, the metric can be written in the form

$$ds^2 = -\,e^{2\nu}(dt)^2 + r^2\sin^2\theta\,(d\varphi)^2 + e^{2\mu_2}\,(dr)^2 + r^2(d\theta)^2\,, \tag{1}$$

where ν and μ_2 are, in the first instance, functions of r and t. And when we restrict ourselves to vacuum fields, as we shall, Birkhoff's theorem assures us that ν and μ_2 are functions of r only.

For the Schwarzschild solution

$$e^{2\nu} = e^{-2\mu_2} = \frac{\Delta}{r^2}\,, \quad \text{where} \quad \Delta = r^2 - 2Mr\,. \tag{2}$$

The horizon of the black hole occurs at

$$r = 2M\,, \quad \text{where} \quad \Delta = 0\,. \tag{3}$$

As is well known, the space interior to $r = 2M$ is incommunicable to the space outside, a consequence of the fact that $r = 2M$ defines a *null surface*, that is, a surface whose tangent at every point is a null vector.

As we have stated, the problem of the perturbation of a black hole is effectively one of determining how waves of different sorts, incident on the black hole, are affected by, and in turn affect, the black hole. In the context of the Schwarzschild black hole, the case of greatest interest is when the incident waves are gravitational in origin.

From general considerations, one may expect that a fraction of the energy in the incident waves will be irreversibly absorbed by the black hole, while the remaining fraction will be scattered (or, reflected) back to infinity. In other words, it would appear that it may be possible to visualize the black hole as presenting a potential barrier to the oncoming waves. We shall see that this expectation is amply fulfilled (not only in the context of the Schwarzschild black hole but in all other contexts as well). Indeed, it will appear that the problem can be reduced to the very simple one of the penetration of one-dimensional potential barriers with which we are familiar in elementary quantum theory. The basic reason why this reduction to a *strict* one-dimensional problem is possible, is the fact that nothing can emerge from the horizon of the black hole and that, in consequence, the region interior to the horizon is of no relevance to our considerations. This latter fact can be expressed in precise mathematical terms by changing to a variable r_* which translates the horizon to minus infinity. Thus, by letting

$$\frac{d}{dr_*} = \frac{\Delta}{r^2}\frac{d}{dr}, \tag{4}$$

or, explicitly,

$$r_* = r + 2M\, ln\left(\frac{r}{2M} - 1\right) \quad (r > 2M), \tag{5}$$

we have the behaviors

$$r_* \rightarrow r \quad \text{as} \quad r \rightarrow \infty \quad \text{and} \quad r_* \rightarrow -\infty \quad \text{as} \quad r \rightarrow 2M + 0\,. \tag{6}$$

The permitted range of r_* is the entire internal $(-\infty, +\infty)$, and we are debarred from making any reference to $r < 2M$.

In terms of the variable r_*, the boundary conditions are readily stated: for $r_* \rightarrow +\infty$, we can have both incoming and outgoing waves, since, in addition to the postulated incident waves, we can also have waves reflected back to infinity by the black hole; but for $r_* \rightarrow -\infty$, we can, only, have waves going towards $-\infty$ since no waves can emerge from the interior of the black hole.

We conclude this section by stating that the perturbations of the black-hole solutions of general relativity can be studied from two different standpoints and two different bases: *either* directly for the metric perturbations via the Einstein (or the Einstein-Maxwell) equations linearized about the unperturbed solution, *or* for the Weyl (and the Maxwell) scalars via the equations of the Newman-Penrose formalism. It is found that the two methods supplement one another very effectively. For this reason, we shall begin with a brief account of the two methods of treating the problem in the explicit context of the Schwarzschild black hole.

3. THE METHOD OF THE METRIC PERTURBATIONS

Since we are dealing with the perturbations of a system which is initially spherically symmetric, it is clear that, in the equations governing the perturbations, the independent variables r, θ, and φ can be separated. In particular, the dependence on φ can be separated by seeking solutions with a φ-dependence given by $e^{im\varphi}$ where m is an integer positive, negative, or zero. On the other hand, inasmuch as the solutions with a φ-dependence $e^{im\varphi}$ can be obtained by subjecting the solutions independent of φ (i.e., with $m = 0$) to a rotation, it will suffice to linearize the exact equations, governing nonstationary axisymmetric systems, about the given time-independent spherically symmetric solution.

A form of the metric adequate for treating nonstationary axisymmetric systems in general relativity is given by

$$\begin{aligned} ds^2 = &- e^{2\nu}(dt)^2 + e^{2\psi}(d\varphi - \omega dt - q_2 dx^2 - q_3 dx^3)^2 \\ &+ e^{2\mu_2}(dx^2)^2 + e^{2\mu_3}(dx^3)^2 \,, \end{aligned} \tag{7}$$

where ν, ψ, μ_2, μ_3, ω, q_2, and q_3 are all functions of t and the spatial coordinates x^2 and x^3 (which we may identify with a "radial coordinate" r and an angular coordinate θ). Further, the functions ω, q_2, and q_3 can occur in the field equations only in the combinations

$$q_{2,3} - q_{3,2} \,, \quad \omega_{,2} - q_{2,0} \,, \quad \text{and} \quad \omega_{,3} - q_{3,0} \,, \tag{8}$$

where "0" signifies "t."

We observe that the metric (7) includes the form

$$\begin{aligned} ds^2 = &- e^{2\nu}(dt)^2 + e^{2\psi}(d\varphi - \omega dt)^2 \\ &+ e^{2\mu_2}(dx^2)^2 + e^{2\mu_3}(dx^3)^2 \,, \end{aligned} \tag{9}$$

which is appropriate for describing stationary axisymmetric systems. And when considering stationary systems, the functions ν, ψ, μ_2, μ_3, and ω are functions only of x^2 and x^3; and, moreover, we have the gauge freedom to impose any coordinate condition on μ_2 and μ_3 which we may choose.

The function ω, which occurs in the metrics (7) and (9), plays a decisive role in distinguishing the character of the perturbations. The physical significance of ω is the following. Let $\Omega(x^2,x^3) = d\varphi/dt$, in the stationary case, denote the angular velocity as measured by an observer at infinity. Then, an observer, who considers himself as locally at rest at (x^2,x^3), will be attributed by an observer at infinity as having an angular velocity $\Omega - \omega$ instead of Ω. This difference in the angular velocities, as perceived by the two observers, is commonly ascribed to the "dragging of the inertial frames" by the angular momentum resident in the field: the angular momentum, J, is in fact determined by the asymptotic behavior, $2J/r^3$, of ω.

Now comparing the metrics (1) and (7), we observe that the Schwarzschild solution ascribes to the various metric functions in (9) the values,

$$e^{2\nu} = e^{-2\mu_2} = \Delta r^{-2} \,, \quad e^{\psi} = r\sin\theta \,, \quad e^{\mu_3} = r \,, \tag{10}$$

and

$$\omega = q_2 = q_3 = 0\,, \tag{11}$$

where, it may be noted that, besides the identifications

$$x^0 = t\,, \quad x^2 = r\,, \quad \text{and} \quad x^3 = \theta\,, \tag{12}$$

the available gauge freedom has been utilized in the choice of the radial coordinate r as defining a "luminosity distance" (which preserves the value $4\pi r^2$ for the area of the two-surface r = constant).

From our earlier remarks, it would appear that we can distinguish among the perturbations of a spherically symmetric solution, two noncombining classes:

(i) Perturbations in which ν, ψ, μ_2, and μ_3 are left unchanged; and perturbation consists only in making ω, q_2, and q_3 nonvanishing (but infinitesimally small so that only terms linear in these quantities need be retained).

(ii) Perturbations in which ν, ψ, μ_2, and μ_3 are subjected to infinitesimal changes while ω, q_2, and q_3 continue to be vanishing.

The perturbations belonging to the two classes are generally referred to as of "odd" and of "even" parities. While the parities of the two classes of perturbations are clearly opposite, the nomenclature "odd" and "even" is unfortunate and misleading. We shall avoid using it.

Equations governing the two classes of perturbations can be obtained by linearizing the Einstein equations appropriate to the metric (7) about the Schwarzschild solution. We shall consider the results of this linearization in § 5 below.

4. THE METHOD BASED ON THE NEWMAN-PENROSE FORMALISM

In the Newman-Penrose formalism, the description of the spacetime is in terms of a local tetrad frame based on four null vectors, $(\mathbf{l}, \mathbf{n}, \mathbf{m}, \overline{\mathbf{m}})$, of which $\mathbf{l}$ and $\mathbf{n}$ are real and $\mathbf{m}$ and $\overline{\mathbf{m}}$ are complex conjugates. It is further supposed that the chosen vectors satisfy the orthogonality relations,

$$\begin{aligned} &\mathbf{l}\cdot\mathbf{n} = 1\,, \quad \mathbf{m}\cdot\overline{\mathbf{m}} = -1\,, \quad \text{and} \\ &\mathbf{l}\cdot\mathbf{m} = \mathbf{l}\cdot\overline{\mathbf{m}} = \mathbf{n}\cdot\mathbf{m} = \mathbf{n}\cdot\overline{\mathbf{m}} = 0\,. \end{aligned} \tag{13}$$

And any tensor (in spacetime), in which we may be interested, is projected onto the chosen frame by contracting it with the basis vectors in all possible nontrivial ways. Thus, in place of the Weyl tensor C_{pqrs}, we consider the five complex scalars

$$\Psi_0 = -\,C_{pqrs}l^p m^q l^r m^s\,,$$

$$\Psi_1 = -\,C_{pqrs}l^p n^q l^r m^s\,,$$

$$\Psi_2 = - C_{pqrs} l^p m^q \bar{m}^r n^s , \tag{14}$$
$$\Psi_3 = - C_{pqrs} l^p n^q \bar{m}^r n^s ,$$
$$\Psi_4 = - C_{pqrs} n^p \bar{m}^q n^r \bar{m}^s ;$$

and in place of the Ricci tensor, we consider the scalars

$$\Phi_{00} = -\frac{1}{2} R_{pq} l^p l^q , \qquad \Phi_{01} = -\frac{1}{2} R_{pq} l^p m^q ,$$
$$\Phi_{11} = -\frac{1}{4} R_{pq} (l^p n^q + m^p \bar{m}^q) , \quad \Phi_{12} = -\frac{1}{2} R_{pq} n^p m^q , \tag{15}$$
$$\Phi_{22} = -\frac{1}{2} R_{pq} n^p n^q , \qquad \Phi_{02} = -\frac{1}{2} R_{pq} m^p m^q .$$

The various equations governing the gravitational and other fields in which we may be interested are similarly projected onto the chosen frame and expressed in terms of the various scalars. It is clear that these equations will involve the covariant derivatives of the basis vectors, also, projected onto the null frame. These are the so-called *spin coefficients*; they are equivalent to the Ricci rotation-coefficients when the chosen frame is an orthonormal one consisting of one timelike and three spacelike vectors. In the Newman-Penrose formalism, the spin coefficients are designated by special symbols:

$$\kappa = m^i l_{i;j} l^j = - l^i m_{i;j} l^j ,$$
$$\sigma = m^i l_{i;j} m^j = - l^i m_{i;j} m^j ,$$
$$\lambda = n^i \bar{m}_{i;j} \bar{m}^j = - \bar{m}^i n_{i;j} \bar{m}^j ,$$
$$\nu = n^i \bar{m}_{i;j} n^j = - \bar{m}^i n_{i;j} n^j ,$$
$$\rho = m^i l_{i;j} \bar{m}^j = - l^i m_{i;j} \bar{m}^j ,$$
$$\mu = n^i \bar{m}_{i;j} m^j = - \bar{m}^i n_{i;j} m^j , \tag{16}$$
$$\tau = m^i l_{i;j} n^j = - l^i m_{i;j} n^j ,$$
$$\pi = n^i \bar{m}_{i;j} l^j = - \bar{m}^i n_{i;j} l^j ,$$
$$\varepsilon = \frac{1}{2} (n^i l_{i;j} l^j + m^i \bar{m}_{i;j} l^j) ,$$
$$\gamma = \frac{1}{2} (n^i l_{i;j} n^j + m^i \bar{m}_{i;j} n^j) ,$$
$$\alpha = \frac{1}{2} (n^i l_{i;j} \bar{m}^j + m^i \bar{m}_{i;j} \bar{m}^j) ,$$

and

$$\beta = \frac{1}{2}\,(n^i l_{i;j} m^j + m^i \overline{m}_{i;j} m^j)\,.$$

In the Newman-Penrose formalism (as indeed, in any tetrad formalism) one writes down three sets of equations: the *Bianchi identities*,

$$R_{ij[kl;m]} = 0 \tag{17}$$

expressed in terms of the spin coefficients and the Weyl and the Ricci scalars; the *commutation relations*,

$$[\mathbf{l}^{(a)}, \mathbf{l}^{(b)}] = C^{(a)(b)}{}_{(c)}\mathbf{l}^{(c)}\,, \tag{18}$$

where we have written $\mathbf{l}^{(1)}$, $\mathbf{l}^{(2)}$, $\mathbf{l}^{(3)}$, and $\mathbf{l}^{(4)}$ in place of $\mathbf{l}$, $\mathbf{n}$, $\mathbf{m}$, and $\overline{\mathbf{m}}$, and the basis vectors are to be interpreted as directional derivatives, and the *structure constants* $C^{(a)(b)}{}_{(c)}$ are expressed in terms of the spin coefficients; and, finally, the components of the Riemann tensor derived from the *Ricci identity*,

$$\mathbf{l}^{(a)}{}_{i;k;l} - \mathbf{l}^{(a)}{}_{i;l;k} = \mathbf{l}^{(a)m} R_{mikl}\,, \tag{19}$$

and written in terms of the spin coefficients, their directional derivatives, and the Weyl and the Ricci scalars.

The Newman-Penrose formalism has had spectacular successes in treating the perturbations of the black-hole solutions of general relativity. The basic reason for this success can be traced to the circumstance that the geometry of the underlying spacetimes (by virtue of their belonging to the so-called "type-D" class) enables a choice of basis, $\mathbf{l}$, $\mathbf{n}$, $\mathbf{m}$, and $\overline{\mathbf{m}}$, such that the vectors $\mathbf{l}$ and $\mathbf{n}$ form shear-free congruences of null geodesics with the consequence that the Weyl scalars Ψ_0, Ψ_1, Ψ_3, and Ψ_4 and the spin coefficients κ, σ, λ, and ν all vanish. These latter facts entail that we can determine these important quantities, when they are nonvanishing in the perturbed state, without solving for the metric perturbation or, indeed, any other quantity.

The Description of the Schwarzschild Geometry in the Newman-Penrose Formalism

A basic null-tetrad satisfying the requirements

$$\mathbf{l}\cdot\mathbf{n} = 1\,, \quad \mathbf{m}\cdot\overline{\mathbf{m}} = -1\,, \quad \text{and} \quad \kappa = \sigma = \lambda = \nu = 0 \tag{20}$$

and appropriate to the Schwarzschild metric is given by

$$l^i = \frac{1}{\Delta}\,(r^2, +\Delta, 0, 0) \qquad ; \quad l_i = \left(\,1, -\frac{r^2}{\Delta}, 0, 0\,\right)\,,$$

$$n^i = \frac{1}{2r^2}(r^2,-\Delta,0,0) \quad ; \quad n_i = \frac{1}{2r^2}(\Delta,+r^2,0,0)\,, \tag{21}$$

$$m^i = \frac{1}{r\sqrt{2}}(0,0,1,i\ \text{cosec}\theta)\,; \quad m_i = \frac{1}{r\sqrt{2}}(0,0,-r^2,-ir^2\sin\theta)\,.$$

The remaining spin coefficients determined by this basis are

$$\rho = -\frac{1}{r}, \quad \beta = \frac{\cot\theta}{r(2\sqrt{2}\,)} = -\alpha\,, \quad \pi = \tau = \varepsilon = 0\,,$$

$$\mu = -\frac{\Delta}{2r^3}, \quad \text{and} \quad \gamma = \mu + \frac{r-M}{2r^2}\,. \tag{22}$$

And it can be verified that, with respect to the chosen basis,

$$\Psi_0 = \Psi_1 = \Psi_3 = \Psi_4 = 0 \tag{23}$$

and

$$\Psi_2 = -Mr^{-3}\,. \tag{24}$$

This completes the description of the Schwarzschild geometry in the Newman-Penrose formalism.

5. THE METRIC PERTURBATIONS OF THE SCHWARZSCHILD BLACK HOLE AND THE POTENTIAL BARRIERS SURROUNDING IT

As we have explained in § 3, we can distinguish two classes of perturbations of opposite parity: those of class I in which ν, ψ, μ_2, and μ_3 remain unchanged, while ω, q_2, and q_3 are nonvanishing, and those of class II in which ν, Ψ, μ_2, and μ_3 experience infinitesimal changes, while ω, q_2, and q_3 remain vanishing.

Considering first the perturbations of class I, we find that with the substitution

$$re^{2\nu}(q_{2,3} - q_{3,2})\sin^3\theta = Z^{(-)}(r)P_{l+2}(\cos\theta \mid -3)e^{i\sigma t}\,, \tag{25}$$

where σ is a constant and $P_{l+2}(\cos\theta \mid -3)$ is Gegenbauer's polynomial of order $l+2$ and index -3, $Z^{(-)}(r)$ satisfies the one-dimensional wave equation

$$\left(\frac{d^2}{dr_*^2} + \sigma^2\right) Z^{(-)} = V^{(-)}Z^{(-)}\,, \tag{26}$$

where

$$V^{(-)} = 2\frac{\Delta}{r^5}[(n+1)r - 3M] \tag{27}$$

and

$$n = \frac{1}{2}(l-1)(l+2) = \mu^2 \text{ (say)}. \tag{28}$$

Equation (26) is generally referred to as the Regge-Wheeler equation.

Turning next to the perturbations of class II, we find that the variables can be separated by the substitutions (due to J. Friedman)

$$\begin{aligned} \delta\nu &= N(r)P_l(\cos\theta)e^{i\sigma t}, \\ \delta\mu_2 &= L(r)P_l(\cos\theta)e^{i\sigma t}, \\ \delta\mu_3 &= [T(r)P_l + V(r)P_{l,\theta\theta}]e^{i\sigma t}, \end{aligned} \tag{29}$$

and

$$\delta\psi = [T(r)P_l + V(r)P_{l,\theta}\cot\theta]e^{i\sigma t},$$

where N, L, T, and V are four radial functions. And we find that Einstein's equations, appropriate to the metric (7), linearized about Schwarzschild's solution, lead to the algebraic relation,

$$T - V + L = 0, \tag{30}$$

and three additional equations expressing the radial derivatives of N, L, and V as linear combinations of these same functions. Making use of these equations, we find that the function,

$$Z^{(+)} = \frac{r^2}{nr+3M}\left(\frac{3M}{r}V - L\right), \tag{31}$$

satisfies the one-dimensional wave equation

$$\left(\frac{d^2}{dr_*^2} + \sigma^2\right) Z^{(+)} = V^{(+)}Z^{(+)}, \tag{32}$$

where

$$V^{(+)} = 2\frac{\Delta}{r^5}\frac{n^2(n+1)r^3 + 3n^2Mr^2 + 9nM^2r + 9M^3}{(nr+3M)^2}. \tag{33}$$

Equation (33) is generally referred to as the Zerilli equation.

The potentials $V^{(\pm)}$ are positive for all values of r_* (see Figures 6.1 and 6.2) and have similar asymptotic behaviors:

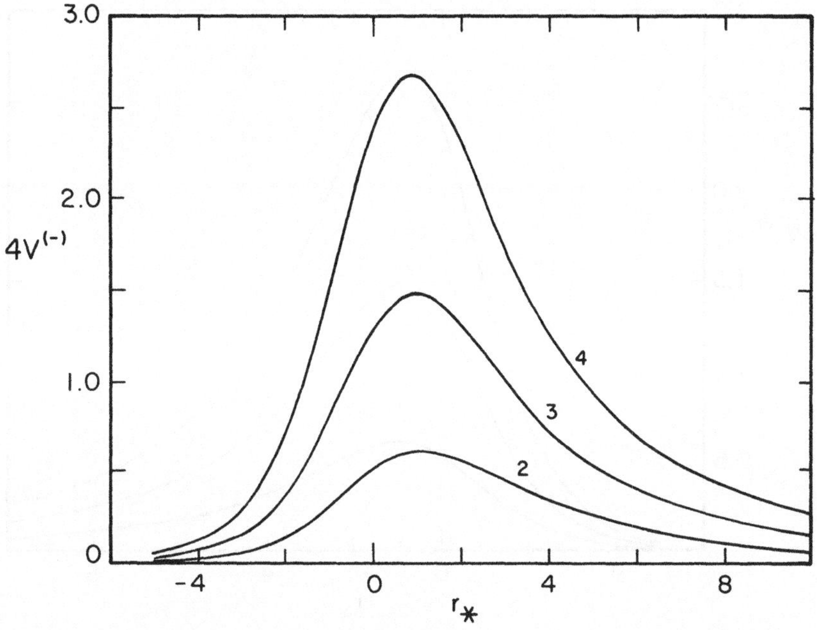

Figure 6.1. The potential barrier $V^{(-)}$ surrounding the Schwarzschild black hole. The curves are labeled by the values of l to which they belong; also r_* is measured in the unit $2M$.

$$\left.\begin{aligned} &V^{(\pm)} \to 2(n+1)r^{-2} && (r \to \infty; r_* \to +\infty) \\ \text{and} \\ &V^{(\pm)} \to (\text{constant})_{\pm}\, e^{r_*/2M} && (r_* \to -\infty)\,. \end{aligned}\right\} \tag{34}$$

The potentials are, therefore, of *short range* (i.e., their integrals over the range of r_* are finite); indeed, it can be verified that

$$2M \int_{-\infty}^{+\infty} V^{(\pm)}\, dr_* = 2n + \frac{1}{2}. \tag{35}$$

This equality of the integrals of $V^{(+)}$ and $V^{(-)}$ over the range of r_* is the manifestation of an important property of the two classes of perturbations to which we shall return in § 10 below.

The Problem of Reflection and Transmission

From the fact that $Z^{(+)}$ and $Z^{(-)}$ satisfy one-dimensional wave equations, we conclude that the normal modes of the black hole are determined by the

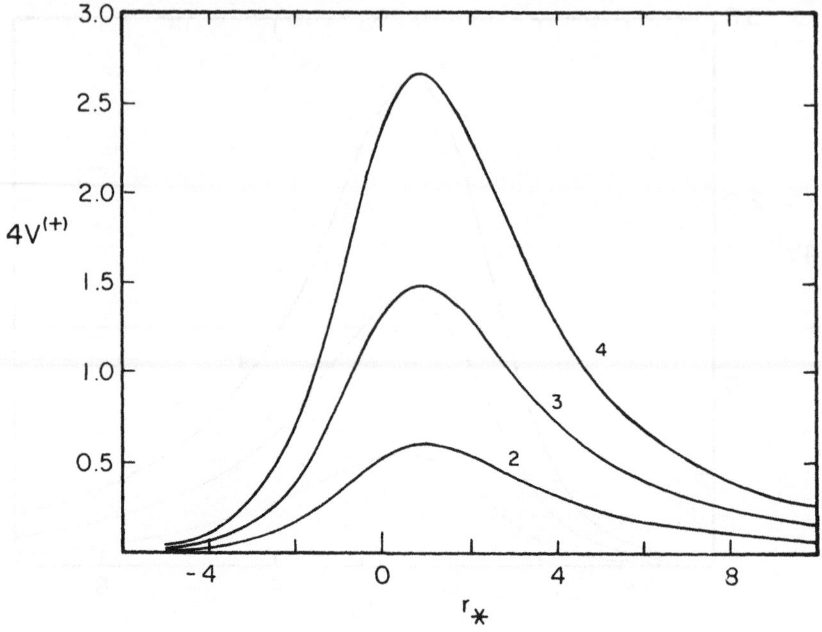

Figure 6.2. The potential barrier $V^{(+)}$ surrounding the Schwarzschild black hole. The curves are labeled by the values of l to which they belong; also r_* is measured in the unit $2M$.

solution to the elementary problem of the penetration of the potential barriers associated with $V^{(+)}$ and $V^{(-)}$. And since the potentials are of short range, they are determined by the solutions of the equations which satisfy the boundary conditions

$$\begin{aligned} Z^{(\pm)} &\rightarrow e^{+i\sigma r_*} + A^{(\pm)}e^{-i\sigma r_*} \quad (r_* \rightarrow +\infty) \\ &\rightarrow \qquad B^{(\pm)}e^{+i\sigma r_*} \quad (r_* \rightarrow -\infty)\,, \end{aligned} \tag{36}$$

where $A^{(\pm)}$ and $B^{(\pm)}$ are the amplitudes of the waves which are reflected and transmitted by the potential barriers, $V^{(\pm)}$, when a wave of unit amplitude and frequency σ is incident on the black hole from infinity (see Figure 6.3). The fractions of the incident flux of energy that are reflected and transmitted are, therefore, given by

$$\mathrm{R}^{(\pm)} = |A^{(\pm)}|^2 \quad \text{and} \quad \mathrm{T}^{(\pm)} = |B^{(\pm)}|^2\,. \tag{37}$$

The quantities $\mathrm{R}^{(\pm)}$ and $\mathrm{T}^{(\pm)}$ define the *reflection* and the *transmission* coefficients appropriate to the two classes of perturbations.

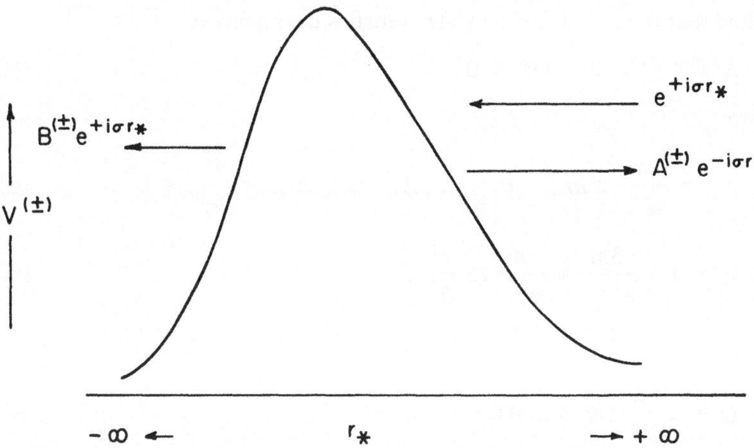

Figure 6.3. Illustrating the problem of reflection and transmission by potential barriers.

From the constancy of the Wronskian of $Z^{(\pm)}$ and their complex conjugates, it follows that, consistent with the meanings of these quantities,

$$R^{(\pm)} + T^{(\pm)} = 1\,. \tag{38}$$

It is a remarkable fact that the reflection and the transmission coefficients for the two classes of perturbations are *identically equal* (i.e., for all values of σ) even though $V^{(+)}$ and $V^{(-)}$ are such entirely different functions. The meaning and the origin of this equality is what we shall be concerned with, principally, in the rest of this article.

6. A WAVE EQUATION WHICH FOLLOWS FROM THE NEWMAN-PENROSE FORMALISM

The equations of the Newman-Penrose formalism provide equations which are already linearized in the sense that they are linear and homogeneous in the quantities which vanish in the unperturbed state, namely, Ψ_0, Ψ_1, Ψ_3, Ψ_4, κ, σ, λ, and ν. These quantities can, therefore, be determined without any knowledge of how the perturbation affects any of the other quantities. The reduction of these equations is fairly straightforward and leads to wave equations for Ψ_0 and Ψ_4. Thus, writing

$$\Psi_0 = \frac{r^3}{2\Delta^2}\,Y(r)(P_{l,\theta\theta} - P_{l,\theta}\cot\theta)e^{i\sigma t}\,, \tag{39}$$

we find that the radial function $Y(r)$ satisfies the equation

$$\Lambda^2 Y + P\Lambda_- Y - QY = 0\,, \tag{40}$$

where

$$\Lambda_\pm = \frac{d}{dr_*} \pm i\sigma\,, \quad \Lambda^2 = \Lambda_+\Lambda_- = \Lambda_-\Lambda_+ = \frac{d^2}{dr_*^2} + \sigma^2\,, \tag{41}$$

$$P = 4\,\frac{r-3M}{r^2} = \frac{d}{dr_*}\,lg\,\frac{r^8}{\Delta^2}\,, \tag{42}$$

and

$$Q = 2\,\frac{\Delta}{r^5}\,(nr + 3M)\,. \tag{43}$$

A similar consideration of Ψ_4 leads to the complex conjugate of equation (40).

It can be shown that the solutions for the perturbations in all the remaining quantities can be expressed in terms of the solution for Y. Accordingly, it would appear that the functions $Z^{(\pm)}$ must also be related to Y. We consider in the following sections how these relations are to be found.

7. THE TRANSFORMATION THEORY

Our aim is to transform equation (40) to an equation of the form,

$$\Lambda^2 Z = VZ\,, \tag{44}$$

where V is some function of r_*(or r) to be determined. It is clear that Z (on the assumption that one such exists) must be a linear combination of Y and its derivative. There is, therefore, no loss of generality in assuming that Z is related to Y in the manner,

$$Y = f\,VZ + (W + 2i\sigma f\,)\Lambda_+ Z\,, \tag{45}$$

where f, V, and W are certain functions unspecified for the present. Applying the operator Λ_- to equation (45) and making use of the fact that Z has been assumed to satisfy equation (44), we find that

$$\Lambda_- Y = -\,\beta\,\frac{\Delta^2}{r^8}\,Z + R\Lambda_+ Z\,. \tag{46}$$

where

$$-\,\frac{\Delta^2}{r^8}\,\beta = \frac{d}{dr_*}(f\,V) + WV \tag{47}$$

and

$$R = fV + \frac{d}{dr_*}(W + 2i\sigma f). \tag{48}$$

The requirement that equations (45) and (46) are compatible with equation (40), satisfied by Y, leads to the pair of equations,

$$\frac{d}{dr_*}\left(\frac{r^8}{\Delta^2}R\right) = \frac{r^8}{\Delta^2}[Q(W+2i\sigma f) - 2i\sigma R] + \beta \tag{49}$$

and

$$-\frac{\Delta^2}{r^8}\frac{d\beta}{dr_*} = (Qf-R)V. \tag{50}$$

It can be verified that equations (47)–(50) allow the integral

$$\frac{r^8}{\Delta^2}RfV + \beta(W+2i\sigma f) = K = \text{constant}. \tag{51}$$

Accordingly, it will suffice to consider any four of the five equations (47)–(51).

It will be observed that we are provided with only four independent equations for determining the five functions β, f, R, V, and W. We have, therefore, a considerable latitude in seeking solutions of equations (47)–(51).

Dual Transformations and the Conditions for the Existence of Transformations with β = Constant and $f = 1$

We shall presently verify that for the particular cases in which we are interested, equations (47)–(51) allow solutions consistently with the assumptions,

$$f = 1 \quad \text{and} \quad \beta = \text{constant}. \tag{52}$$

These assumptions are clearly over-restrictive if we consider Q as some arbitrarily given function. We are, therefore, essentially asking how Q should be restricted in order that the transformation equations (47)–(51) may allow solutions compatible with the assumptions of (52).

The assumptions (52) in conjunction with equations (50), (48), (49), and (51), successively require

$$R = Q, \qquad V = Q - \frac{dW}{dr_*}, \tag{53}$$

$$\frac{d}{dr_*}\left(\frac{r^8}{\Delta^2}Q\right) = \frac{r^8}{\Delta^2}QW + \beta, \tag{54}$$

and

$$\frac{r^8}{\Delta^2} QV + \beta W = K - 2i\sigma\beta = \kappa \text{ (say)}. \tag{55}$$

(Note that κ is to be a constant.)

With the definition

$$F = r^8 Q/\Delta^2 , \tag{56}$$

equations (54) and (55) give

$$W = \frac{1}{F}\left(\frac{dF}{dr_*} - \beta\right) \tag{57}$$

and

$$FV + \beta W = \kappa . \tag{58}$$

From equations (53) and (58) it now follows that

$$F\left(Q - \frac{dW}{dr_*}\right) + \beta W = \kappa . \tag{59}$$

Finally, eliminating W from equation (59) with the aid of equation (57), we obtain the basic condition,

$$\frac{1}{F}\left(\frac{dF}{dr_*}\right)^2 - \frac{d^2F}{dr_*^2} + \frac{\Delta^2}{r^8} F^2 = \frac{\beta^2}{F} + \kappa . \tag{60}$$

Thus, a necessary and sufficient condition, for the transformation equations (47)–(51) to allow solutions consistently with the assumptions (52), is that equation (60) be satisfied by the given F (i.e., Q) for some suitably chosen constants β^2 and κ. Since β occurs as β^2 in equation (60), it follows that if equation (60) is satisfied for some values of β^2 and κ, then *two* transformations, associated with $+\beta$ and $-\beta$ (but the same κ), are possible. We shall call these *dual transformations.*

8. THE EXPLICIT EXPRESSION OF $Z^{(+)}$ AND $Z^{(-)}$ IN TERMS OF Y

We shall now verify that equations (47)–(51) allow solutions, consistently with the assumptions (52), for the potential

$$V^{(-)} = \frac{\Delta}{r^5}[(\mu^2+2)r - 6M] \quad (\mu^2 = 2n), \tag{61}$$

appropriate for perturbations of class I (cf. equation (27)).

We first observe that the assumptions (52), in conjunction with equation (50), require that (cf. equations (53) and (43))

$$R = Q = \frac{\Delta}{r^5}(\mu^2 r + 6M). \tag{62}$$

From equations (48), (61), and (62) it now follows that

$$\frac{dW^{(-)}}{dr_*} = Q - V^{(-)} = \frac{\Delta}{r^5}(-2r + 12M), \tag{63}$$

or, making use of equation (4),

$$\frac{dW^{(-)}}{dr} = -\frac{2}{r^2} + \frac{12M}{r^3}. \tag{64}$$

The required solution of this equation is

$$W^{(-)} = \frac{2}{r^2}(r - 3M). \tag{65}$$

We next verify that when the expressions (61) and (65) for $V^{(-)}$ and $W^{(-)}$ are substituted in equation (50), we obtain, consistently with the assumptions (52), that

$$\beta^{(-)} = -6M = \text{constant}. \tag{66}$$

And we complete the verification of the entire set of the transformation equations by observing that equation (51) yields the constant value,

$$K^{(-)} = \mu^2(\mu^2+2) - 12i\sigma M. \tag{67}$$

We conclude that equation (40) can be transformed to the one-dimensional wave equation (26) by the substitutions,

$$Y = V^{(-)}Z^{(-)} + (W^{(-)}+2i\sigma)\Lambda_+ Z^{(-)} \tag{68}$$

and

$$\Lambda_- Y = {}_+6M\frac{\Delta^2}{r^8}Z^{(-)} + Q\Lambda_+ Z^{(-)}, \tag{69}$$

where $V^{(-)}$ and $W^{(-)}$ are given by equations (61) and (65).

The inverse of the relations (68) and (69) are

$$\frac{\Delta^2}{r^8}K^{(-)}Z^{(-)} = QY - (W^{(-)}+2i\sigma)\Lambda_- Y \tag{70}$$

and

$$\frac{\Delta^2}{r^8} K^{(-)} \Lambda_+ Z^{(-)} = -6M \frac{\Delta^2}{r^8} Y + V^{(-)} \Lambda_- Y \,. \tag{71}$$

We have shown that if the transformation equations (47)–(51) allow a solution, consistently with the assumptions (52), for some value of $+\beta$, then they must allow a solution also for $-\beta$. Therefore, in the present context, we shall obtain a further transformation associated with the value,

$$\beta^{(+)} = +6M \,. \tag{72}$$

We shall now verify that the resulting *dual transformation* leads to the equation for $Z^{(+)}$.

Now, letting (cf. equation (56))

$$F = \frac{r^8}{\Delta^2} Q = \frac{r^3}{\Delta} (\mu^2 r + 6M) \,, \tag{73}$$

we deduce from equation (57) that

$$W^{(+)} = \frac{d}{dr_*} lgF - \frac{6M}{F} . \tag{74}$$

Since (as may be readily verified)

$$\begin{aligned} \frac{d}{dr_*} lgF &= \frac{2}{r^2}(r - 3M) - \frac{6M\Delta}{r^3(\mu^2 r + 6M)} \\ &= W^{(-)} - \frac{6M\Delta}{r^3(\mu^2 r + 6M)} \,, \end{aligned} \tag{75}$$

it follows that

$$W^{(+)} = W^{(-)} - \frac{12M\Delta}{r^3(\mu^2 r + 6M)} . \tag{76}$$

The corresponding expression for $V^{(+)}$ is given by (cf. equation (48))

$$V^{(+)} = V^{(-)} + 12M \frac{d}{dr_*} \frac{\Delta}{r^3(\mu^2 r + 6M)} . \tag{77}$$

It can be verified that this expression for $V^{(+)}$ is in agreement with that given in equation (33).

From equation (77) it follows that

$$\textstyle\int_{-\infty}^{+\infty} V^{(+)}\, dr_* = \int_{-\infty}^{+\infty} V^{(-)}\, dr_* \,, \tag{78}$$

an equality to which we made reference in § 5 (equation (35)).

Finally, we may note that, in agreement with equation (55), we find that

$$\frac{r^8}{\Delta^2}QV^{(+)} + 6M(W^{(+)}+2i\sigma) = K^{(+)} = \mu^2(\mu^2+2) + 12i\sigma M\,. \quad (79)$$

The transformation equations, appropriate for the perturbations of class II, that are analogous to equations (68)–(71), are

$$Y = V^{(+)}Z^{(+)} + (W^{(+)}+2i\sigma)\Lambda_+Z^{(+)}\,, \quad (80)$$

$$\Lambda_-Y = -\,6M\frac{\Delta^2}{r^8}Z^{(+)} + Q\Lambda_+Z^{(+)}\,, \quad (81)$$

$$\frac{\Delta^2}{r^8}K^{(+)}Z^{(+)} = QY - (W^{(+)}+2i\sigma)\Lambda_-Y\,, \quad (82)$$

and

$$\frac{\Delta^2}{r^8}K^{(+)}\Lambda_+Z^{(+)} = +\,6M\frac{\Delta^2}{r^8}Y + V^{(+)}\Lambda_-Y\,. \quad (83)$$

9. AN EXPLICIT RELATION BETWEEN THE SOLUTIONS FOR $Z^{(+)}$ AND $Z^{(-)}$ AND THE PROOF OF THE EQUALITY OF THE COEFFICIENTS $R^{(+)}$ AND $R^{(-)}$ AND $T^{(+)}$ AND $T^{(-)}$

We have seen how equation (40) for Y can be transformed to the one-dimensional wave equations governing $Z^{(+)}$ and $Z^{(-)}$ by transformations which we have called dual. We shall now show, with the aid of the transformation equations (68)–(71) and (80)–(83), that there is a simple relation which will enable us to obtain a solution $Z^{(-)}$ appropriate to perturbations of class I, from a solution $Z^{(+)}$ appropriate to perturbations of class II, and conversely.

Let $Z^{(-)}$ denote a solution of the wave equation (26). Then making use of equation (70), relating $Z^{(-)}$ to Y and Λ_-Y, and then, of equations (80) and (81), relating Y and Λ_-Y to $Z^{(+)}$ and $\Lambda_+Z^{(+)}$, we obtain,

$$\begin{aligned}\frac{\Delta^2}{r^8}K^{(-)}Z^{(-)} &= QY - (W^{(-)}+2i\sigma)\Lambda_-Y\\ &= Q[V^{(+)}Z^{(+)} + (W^{(+)}+2i\sigma)\Lambda_+Z^{(+)}]\\ &\quad - (W^{(-)}+2i\sigma)\left[-6M\frac{\Delta^2}{r^8}Z^{(+)} + Q\Lambda_+Z^{(+)}\right], \quad (84)\end{aligned}$$

or, after some regrouping of the terms, we have

$$\frac{\Delta^2}{r^8}K^{(-)}Z^{(-)} = \frac{\Delta^2}{r^8}\left[\frac{r^8}{\Delta^2}QV^{(+)} + 6M(W^{(+)}+2i\sigma)\right.$$

$$+ 6M(W^{(-)}-W^{(+)})\Big]Z^{(+)}$$

$$+ Q(W^{(+)}-W^{(-)})\Lambda_+ Z^{(+)}. \tag{85}$$

The expression on the right-hand side of this equation can be simplified by making use of equations (67), (76), and (79). We find

$$[\mu^2(\mu^2+2) - 12i\sigma M]Z^{(-)} = \left[\mu^2(\mu^2+2) + \frac{72M^2\Delta}{r^3(\mu^2 r+6M)}\right]Z^{(+)}$$

$$- 12M\frac{dZ^{(+)}}{dr_*}. \tag{86}$$

The analogous equation relating $Z^{(+)}$ to $Z^{(-)}$ is

$$[\mu^2(\mu^2+2) + 12i\sigma M]Z^{(+)} = \left[\mu^2(\mu^2+2) + \frac{72M^2\Delta}{r^3(\mu^2 r+6M)}\right]Z^{(-)}$$

$$+ 12M\frac{dZ^{(-)}}{dr_*}. \tag{87}$$

Equations (86) and (87) clearly enable us to derive a solution (appropriate for describing perturbations of class I) from a solution (appropriate for describing perturbations of class II), and conversely.

Since the second term in square brackets on the right-hand sides of equations (86) and (87) vanishes on the horizon ($r_* \to -\infty$ and $\Delta = 0$) and at infinity ($r_* \to +\infty$), it follows that

$$[\mu^2(\mu^2+2) \pm 12i\sigma M]Z^{(\pm)} \to \mu^2(\mu^2+2)Z^{(\mp)} \pm 12M\frac{dZ^{(\mp)}}{dr_*} \qquad (r_* \to \pm\infty). \tag{88}$$

In particular, solutions for $Z^{(+)}$ derived from solutions for $Z^{(-)}$ (in accordance with equation (87)) having the asymptotic behaviors,

$$Z^{(-)} \to e^{+i\sigma r_*} \quad \text{and} \quad Z^{(-)} \to e^{-i\sigma r_*} \; (r_* \to \pm\infty), \tag{89}$$

have, respectively, the asymptotic behaviors,

$$Z^{(+)} \to e^{+i\sigma r_*} \text{ and } Z^{(+)} \to \frac{\mu^2(\mu^2+2) - 12i\sigma M}{\mu^2(\mu^2+2) + 12i\sigma M}e^{-i\sigma r_*} \qquad (r_* \to \pm\infty). \tag{90}$$

From these behaviors, it follows that, in equations (36),

$$A^{(+)} = A^{(-)}e^{i\delta} \text{ and } B^{(+)} = B^{(-)}, \tag{91}$$

where

$$e^{i\delta} = \frac{\mu^2(\mu^2+2) - 12i\sigma M}{\mu^2(\mu^2+2) + 12i\sigma M}; \tag{92}$$

and, therefore,

$$|A^{(+)}|^2 = |A^{(-)}|^2 \text{ and } |B^{(+)}|^2 = |B^{(-)}|^2. \tag{93}$$

These relations establish the equality of the reflection and the transmission coefficients for the perturbations belonging to the two classes. It should, however, be noted that, while there is a difference in the relative phases of the reflected amplitudes, there is no such difference in the transmitted amplitudes: they are, in fact, *identically equal.*

10. NECESSARY CONDITIONS FOR DIFFERENT POTENTIALS TO YIELD THE SAME TRANSMISSION AMPLITUDES

The remarkably simple relation between the solutions for $Z^{(+)}$ and $Z^{(-)}$ and the equality of the derived reflection and transmission coefficients suggest the consideration of the following problem in the theory of "inverse scattering."

The one-dimensional wave equation,

$$\frac{d^2\psi}{dx^2} + (\sigma^2 - V)\psi = 0 \quad (-\infty < x < +\infty), \tag{94}$$

where $V(x)$ is a potential whose integral over the range of x is bounded, allows solutions with the asymptotic behaviors,

$$\left.\begin{aligned} \psi \rightarrow e^{+i\sigma x} + A(\sigma)e^{-i\sigma x} \quad (x \rightarrow +\infty) \\ \rightarrow \qquad B(\sigma)e^{+i\sigma x} \quad (x \rightarrow -\infty)\,. \end{aligned}\right\} \tag{95}$$

We now ask for the conditions on V which will yield the same complex amplitude $B(\sigma)$. This problem is more general than the celebrated one of establishing the uniqueness (or, otherwise) of V from a knowledge of the "S-matrix" (i.e., from a knowledge of *both* $A(\sigma)$ and $B(\sigma)$). In considering this latter problem, it is customary to restrict oneself to potentials $V(x)$ which satisfy the requirement,

$$\int_{-\infty}^{+\infty} (1 + |x|)V(x)\, dx \tag{96}$$

is bounded.

When $V(x)$ satisfies this requirement, it has been shown that the amplitude $B(\sigma)$, for complex σ, is an analytic function of σ in the lower half of the complex plane. The potentials in which we are interested have an inverse square behavior for $x \to +\infty$ and, therefore, do not satisfy the requirement (96). We shall, nevertheless, assume that $B(\sigma)$ allows a Laurent expansion for $Im\sigma < 0$—an assumption whose validity remains to be established.

Writing

$$\psi = e^{i\sigma x + w}, \tag{97}$$

we find that w satisfies the differential equation,

$$w_{,xx} + 2i\sigma w_{,x} + (w_{,x})^2 - V = 0\,. \tag{98}$$

By the further substitutions,

$$w = -\int_x^\infty v(x^1,\sigma)dx^1 \quad \text{and} \quad w_{,x} = -v\,, \tag{99}$$

we obtain the Riccati equation,

$$v_{,x} + 2i\sigma v + v^2 - V = 0\,. \tag{100}$$

The boundary conditions (95) imposed on ψ, now require that

$$\left.\begin{aligned} w &\to ln\,B \quad \text{for} \quad x \to -\infty \\ &\to 0 \quad\;\; \text{for} \quad x \to +\infty \end{aligned}\right\} (Im\sigma < 0)\,; \tag{101}$$

and, also, that

$$ln\,B = -\int_{-\infty}^{+\infty} v(x,\sigma)dx\,. \tag{102}$$

And, finally, it follows from equation (100) that

$$v \to \frac{V}{2i\sigma} \quad \text{as} \quad \sigma \to \infty\,. \tag{103}$$

We now suppose that v can be expanded in a Laurent series of the form,

$$v = \sum_{n=1}^{\infty} \frac{v_n(x)}{(2i\sigma)^n}, \tag{104}$$

where, according to equation (103),

$$v_1 = V\,. \tag{105}$$

Inserting the series expansion (104) in equation (102), we obtain

$$lnB = -\sum_{n=1}^{\infty} \frac{1}{(2i\sigma)^n} \int_{-\infty}^{+\infty} v_n(x)dx\,, \tag{106}$$

where we have assumed that integration, term by term, is permissible. Thus, we have an expansion for lnB of the form

$$lnB = -\sum_{n=1}^{\infty} c_n\sigma^{-n}\,, \tag{107}$$

where

$$(2i)^n c_n = \int_{-\infty}^{+\infty} v_n dx . \tag{108}$$

Now inserting the expansion (104) in equation (100) and assuming that we can equate the coefficients of the different inverse powers of σ, we obtain the recurrence relation

$$v_n = -\frac{dv_{n-1}}{dx} - \sum_{l=1}^{n-2} v_l v_{n-1-l} . \tag{109}$$

Using this recurrence relation, we can solve for the v_n's, successively, starting with (cf. equation (105))

$$v_1 = V . \tag{110}$$

We find

$$\begin{aligned} v_2 &= -\frac{dv_1}{dx} = -\frac{dV}{dx}, \\ v_3 &= -\frac{dv_2}{dx} - v_1^2 = \frac{d^2V}{dx^2} - V^2, \\ v_4 &= -\frac{dv_3}{dx} - 2v_1 v_2 = -\frac{d^3V}{dx^3} + 2\frac{dV^2}{dx}, \\ v_5 &= -\frac{dV_4}{dx} - 2v_1 v_3 - v_2^2 = \frac{d^4V}{dx^4} - 3\frac{d^2V^2}{dx^2} \\ &\qquad + \left(\frac{dV}{dx}\right)^2 + 2V^3 , \end{aligned} \tag{111}$$

etc.

The coefficients of odd order, c_{2n+1}, in the expansion for *ln B* are, therefore, given by

$$\begin{aligned} &2ic_1 = \int_{-\infty}^{+\infty} V dx ; \quad (2i)^3 c_3 = \int_{-\infty}^{+\infty} V^2 dx ; \\ &(2i)^5 c_5 = \int_{-\infty}^{+\infty} (2V^3 + V_{,x}^2) dx ; \quad \text{etc.,} \end{aligned} \tag{112}$$

while all the coefficients of even order, c_{2n}, vanish.

It would follow from the foregoing, that if different potentials yield the same transmission amplitude B, then the Laurent expansions for *ln B* must coincide and that, therefore, the integrals,

$$\int_{-\infty}^{+\infty} V dx , \quad \int_{-\infty}^{+\infty} V^2 dx , \quad \int_{-\infty}^{+\infty} (2V^3 + V_{,x}^2) dx , \quad \text{etc.,} \tag{113}$$

must be the same when extended over the different potentials.

We have already verified the equality of the first of the integrals in (113) for $V^{(+)}$ and $V^{(-)}$. The equality of the other two integrals (explicitly listed) in (113) can also be verified directly. But we have, as yet, no general proof that the infinity of the integral relations, which follow from the equality of all the c_{2n+1}'s, are satisfied for $V^{(+)}$ and $V^{(-)}$, unless it be that the present analysis may be considered as having established it!

11. RELATION TO THE THEORY OF THE SOLITONS AND THE THEORY OF THE KORTEWEG-DEVRIES EQUATION

The problem considered in the preceding section is related to the classical theory of one-dimensional solitary waves as described in terms of the Korteweg-Devries equation

$$u_{,t} - 6uu_{,x} + u_{,xxx} = 0 . \tag{114}$$

As is well known, this equation allows an infinity of conservation laws. Thus, as may be readily verified, the following identities hold by virtue of equation (114):

$$\begin{aligned} u_{,t} &= (3u^2 - u_{,xx})_{,x} , \\ \left(\frac{1}{2}u^2\right)_{,t} &= \left(2u^3 + \frac{1}{2}u^2_{,x} - uu_{,xx}\right)_{,x} , \\ \left(u^3 + \frac{1}{2}u^2_{,x}\right)_{,t} &= \left(\frac{18}{4}u^4 + 6uu^2_{,x} - 3u^2u_{,xx} \right. \\ &\qquad \left. + \frac{1}{2}u^2_{,xx} - u_{,x}u_{,xxx}\right)_{,x} , \quad \text{etc.} \end{aligned} \tag{115}$$

Therefore,

$$u , \quad u^2 , \quad 2u^3 + u^2_{,x} , \quad \text{etc.,} \tag{116}$$

are conserved quantities. In particular, if u vanishes sufficiently rapidly for $x \rightarrow \pm\infty$, then

$$\int_{-\infty}^{+\infty} u dx , \quad \int_{-\infty}^{+\infty} u^2 dx , \quad \int_{-\infty}^{+\infty} (2u^3 + u^2_{,x}) dx , \quad \text{etc.,} \tag{117}$$

are constants of the motion. We recognize that the foregoing integrals over u are the same as the integrals over V which we encountered in § 10 (equation (113)). The appearance of the same integrals, in the problem considered in § 10 and in the present context of the Korteweg-Devries equation, is not surprising since the solutions of the latter equation are often expressed in terms of the theory of inverse scattering.

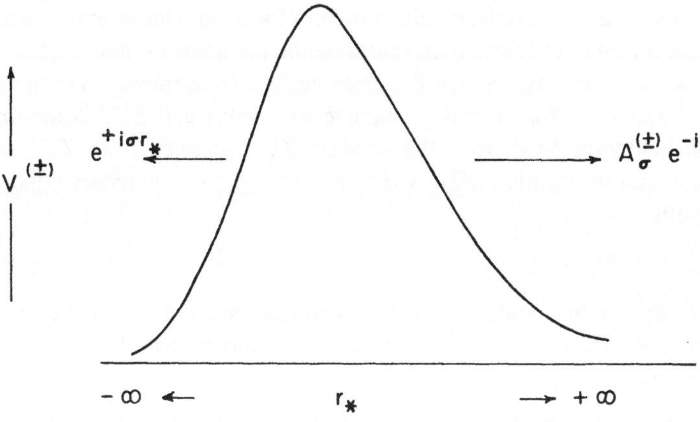

Figure 6.4. Illustrating the nature of the quasi-normal modes.

12. THE QUASI-NORMAL MODES

One may conceive of a black hole being perturbed in a variety of ways: by some object falling into it or by the accretion of matter surrounding it. Or, one may conceive of a black hole being formed by a slightly aspherical collapse of a star and settling toward a final state described by the Schwarzschild solution. In all these cases, the evolution of the initial perturbation—if it can be considered as small—can, in principle, be followed by expressing it as a superposition of the normal modes. However, we may expect that any initial perturbation will, during its last stages, decay in a manner characteristic of the black hole itself and independently of the cause. In other words, we may expect that during these last stages, the black hole emits gravitational waves with frequencies and rates of damping that are characteristic of the black hole itself, in the manner of a bell sounding its last dying notes. These considerations motivate the formulation of the concept of the *quasi-normal modes* of a black hole.

Precisely, quasi-normal modes are defined as proper solutions of the perturbation equations belonging to complex characteristic frequencies which are appropriate for purely ingoing waves at the horizon and purely outgoing waves at infinity (see Figure 6.4).

The problem then is to seek solutions of the equations governing $Z^{\pm}$ which satisfy the boundary conditions,

$$\left.\begin{aligned} Z^{(\pm)} &\to A_\sigma^{(\pm)} e^{-i\sigma r_*} \quad (r_* \to +\infty) \\ &\to \quad e^{+i\sigma r_*} \quad (r_* \to -\infty)\,. \end{aligned}\right\} \tag{118}$$

This is clearly a characteristic value problem for σ, and the solutions belonging to the different characteristic values define the quasi-normal modes.

First, we may observe that the characteristic frequencies σ are the same for $Z^{(+)}$ and $Z^{(-)}$: for, if σ is a characteristic value and $Z_\sigma^{(-)}$ is the proper solution belonging to it, then the solution $Z_\sigma^{(+)}$, derived from $Z_\sigma^{(-)}$ in accordance with the relation (87), will satisfy the required boundary conditions (120) with

$$A_\sigma^{(+)} = A_\sigma^{(-)} e^{i\delta} , \tag{119}$$

where δ is given by equation (92). It will suffice then to consider the equation governing $Z^{(-)}$ (since the potential $V^{(-)}$ is simpler than $V^{(+)}$).

Letting

$$Z^{(-)} = \exp\,(i \textstyle\int^x \phi dr_*) , \tag{120}$$

we have to solve the equation,

$$i\phi_{,r_*} + \sigma^2 - \phi^2 - V^{(-)} = 0 , \tag{121}$$

which satisfies the boundary conditions

$$\phi \to -\sigma \text{ as } r_* \to +\infty \text{ and } \phi \to +\sigma \text{ as } r_* \to -\infty , \tag{122}$$

assuming (as we shall find) that the real part of σ is positive. Solutions having these properties (generally) exist when σ assumes one of a discrete set of complex values, but the set need not be an enumerable infinity: it often is, but sometimes it is not.

An identity, which follows from integrating equation (121) over the entire range of r_* and making use of the boundary condition (122), is

$$-\,2i\sigma + \textstyle\int_{-\infty}^{+\infty} (\sigma^2 - \phi^2) dr_* = \int_{-\infty}^{+\infty} V^{(-)} dr_* . \tag{123}$$

By virtue of the boundary conditions (122), the integral on the left-hand side of this equation exists, while the integral on the right-hand side has a known value (cf. equation (35)).

In Table 6.1 we list the characteristic values of σ for different values of l. Detailed calculations pertaining to the aspherical collapse of dust clouds and of particles falling into black holes along geodesics do manifest the phenomenon of their ringing with the characteristic frequencies and rates of damping of their quasi-normal modes.

13. CONCLUDING REMARKS

As we have pointed out at the outset, several of the features we encounter in the study of the perturbations of the Schwarzschild black hole reappear in different guises and in more general contexts when we study the perturbations of

TABLE 6.1. The Complex Characteristic Frequencies Belonging to the Quasi-Normal Modes of the Schwarzschild Black Hole

l^{a}	$2M\sigma^{b}$	l^{a}	$2M\sigma^{b}$
2	$0.74734 + 0.17792i$	4	$1.61835 + 0.18832i$
	$0.69687 + 0.54938i$		$1.59313 + 0.56877i$
3	$1.19889 + 0.18541i$		$1.12019 + 0.84658i$
	$1.16402 + 0.56231i$	5	$2.02458 + 0.18974i$
	$0.85257 + 0.74546i$	6	$2.42402 + 0.19053i$

[a] The entries in the different lines for $l = 2$, 3, and 4 correspond to the characteristic values belonging to different modes.

[b] Here σ is expressed in the unit $(2M)^{-1}$.

the other black hole solutions of general relativity. Thus, the treatment of the perturbations of the Reissner-Nordström black hole proceeds along lines which parallel the treatment of the perturbations of the Schwarzschild black hole which we have outlined: two classes of perturbations, of opposite parities, can again be distinguished and relations between them, analogous to those between $Z^{(+)}$ and $Z^{(-)}$, can be established. Similarly, the study of the perturbations of the Kerr metric leads us to consider novel problems in the theory of barrier penetration involving complex and singular potentials and manifesting super-radiance (when $\mathbf{R} + \mathbf{T} = 1$ has to be replaced by $\mathbf{R} - \mathbf{T} = 1$). But in the consideration of all these various problems, a comprehensive understanding of the theory of the perturbations of the Schwarzschild black hole appears as an essential prerequisite.

BIBLIOGRAPHICAL NOTES

We shall not provide a complete bibliography which is extensive even if restricted to the Schwarzschild solution and its perturbations. We shall give, instead, references only to those articles and papers in which the reader will find details of the derivation which are omitted, or supplementary information which bears directly on what is stated in the text. Besides, the present account is a "personal view" which a scholarly reader will doubtless find as grossly unbalanced.

§3. The various components of the Riemann, the Ricci, and the Einstein tensor, which are appropriate to the nonstationary axisymmetric metric (7), are listed in

Chandrasekhar, S., and Friedman, John L., "On the Stability of Axisymmetric Systems to Axisymmetric Perturbations in General Relativity: I. The Equations Concerning Nonstationary, Stationary, and Perturbed Systems," *Astrophys. J.* **175**, 379–405 (1972).

§4. For elementary accounts of the tetrad and the Newman-Penrose formalisms, see

Landau, L. D., and Lifshitz, E. M., *Classical Theory of Fields* (Pergamon Press, Oxford, 4th ed., 1975), pp. 291–294.

Chandrasekhar, S., "An Introduction to the Kerr Metric and Its Perturbations," in *General Relativity—An Einstein Centenary Survey* (edited by W. Israel and S. W. Hawking, Cambridge University Press, Cambridge, 1979*a*), Chapter 7, pp. 370–453.

§§5–9. The account in these sections is a rearrangement of the material contained in

Friedman, John L., "On the Born Approximation for Perturbations of a Spherical Star and the Newman-Penrose Constants," *Proc. R. Soc. London Ser. A* **335**, 163–190 (1973).

Chandrasekhar, S., "On the Equations Governing the Perturbations of the Schwarzschild Black Hole," *Proc. R. Soc. London Ser. A* **343**, 289–298 (1975).

Chandrasekhar, S., and Detweiler, S., "The Quasi-normal Modes of the Schwarzschild Black Hole," *Proc. R. Soc. London Ser. A* **344**, 441–452 (1975*a*).

———, "On the Equations Governing the Axisymmetric Perturbations of the Kerr Black Hole," *Proc. R. Soc. London Ser. A* **345**, 145–167 (1975*b*).

For an account with a different emphasis, see

Thorne, K. S., "General Relativistic Astrophysics," in *Theoretical Principles in Astrophysics and Relativity* (edited by N. R. Lebovitz, W. H. Reid, and P. Vandervoort, University of Chicago Press, 1978), pp. 149–216.

§10. The treatment in this section is derived from

Faddeev, L. D., "On the Relation between the S-matrix and potential for the one-dimensional Schrodinger Equation," *Tr. Mat. Inst. V. A. Steklova* **73**, 314–336 (1964).

§11. For a readable account of the theory of the solitary waves and the Korteweg–Devries equation, see

Whitham, G. B., *Linear and Nonlinear Waves* (J. Wiley & Sons, New York, 1974), §§16.14–16.16 and 17.2–17.4.

§12. The complex characteristic values of σ, belonging to the quasi-normal modes, listed in Table 1 are taken from Chandrasekhar and Detweiler (1975*a*) (for l = 2, 3, and 4) and D. L. Gunter ("A Study of the Coupled Gravitational and Electromagnetic Perturbations to the Reissner-Nordström Black Hole: The Scattering Matrix, Energy Conversion, and Quasi-normal Modes," *Phil. Trans. Roy. Soc. Lond. A* **296**, 497–526 (1980); for l = 5 and 6).

§13. For a treatment of the perturbations of the Reissner-Nordström black hole which parallels the treatment of the perturbations of the Schwarzschild black hole, see

Chandrasekhar, S., "On the Equations Governing the Perturbations of the Reissner-Nordström Black Hole," *Proc. R. Soc. London Ser. A* **365**, 453–465 (1979*b*).

Chandrasekhar, S., and Xanthopoulos, B., "On the Metric Perturbations of the Reissner-Nordström Black Hole," *Proc. R. Soc. London Ser. A* **367**, 1–14 (1979).

And for an account of the more general problems in the theory of the penetration of one-dimensional potential barriers which arise in the context of the Kerr metric, see Chandrasekhar (1979*a*).

7. The Initial Value Problem and Beyond

JAMES W. YORK, JR., and TSVI PIRAN

1. THE NATURE OF INITIAL VALUE PROBLEMS

The initial value problem of a physical theory is ascertaining what data must be specified at a given time in order that the equations of motion determine uniquely the evolution of the system. The initial data describe the state of the system at a "fixed time," which, in relativistic theories, means "on a three-dimensional spacelike hypersurface." The initial value and evolution problems taken together constitute what is called the Cauchy problem of the theory. The essence of a Cauchy problem is prediction of the dynamical behavior of a system resulting from a given initial state.

Another important kind of initial value problem results when the data are specified at a given retarded or advanced time, that is, on a three-dimensional null hypersurface. Such "characteristic" initial value problems are often quite appropriate, for example, in the study of radiation fields. The physical and mathematical aspects of characteristic problems are different from those of Cauchy problems. We shall not treat them in the present work.

As the simplest example of an ordinary Cauchy problem, consider a system with a finite number N of degrees of freedom described by N generalized coordinates q^i, N conjugate momenta p_i, and a Hamiltonian $H(q, p)$. Hamilton's equations provide the first order equations of motion

$$\dot{q}^i = \frac{\partial H}{\partial p_i}, \quad \dot{p}_i = -\frac{\partial H}{\partial q^i}. \tag{1}$$

The solutions have the form $q^i = Q^i(t; q_0, p_0)$, $p_i = P_i(t; q_0, p_0)$ and are unique for some finite (or possibly infinite) time t if the initial data q_0, p_0 (values of q^i and p_i at $t = t_0$) are given.

Several features of this elementary problem are worth noting for future reference. First, there is no ambiguity in the meaning of time in Newtonian mechanics, where time is absolute up to trivial changes of units or origin that are the same everywhere in space. Second, there are no constraints on the initial data; that is, q_0 and p_0 are freely specifiable. Third, only the one particular time $t = t_0$ is involved in the description of the initial state of the system, which fact results from using first order (in time) equations of motion.

If we describe the above problem using a Lagrangian approach, we obtain second order equations of motion

$$\ddot{q}^i = F^i(q, \dot{q})$$

from the Lagrange equations

$$\frac{\partial L}{\partial q^i} - \frac{d}{dt}\frac{\partial L}{\partial \dot{q}^i} = 0\,. \tag{2}$$

The initial data are now q_0 and $\dot{q}_0$. The explicit presence of velocity variables among the initial data shows that here we must fix the state of the system, in terms of q^i, at two differing times and then take the limit as these times approach coincidence. Because time is absolute in Newtonian mechanics, there is no significant practical difference between using (q_0, p_0) or $(q_0, \dot{q}_0)$ as initial data. However, in relativistic theories, especially those involving curved spacetime, there is a great deal of arbitrariness in the choice of time variables. As a consequence, it turns out that the distinction between data specifiable at one time and that requiring as a matter of principle two times is quite significant. The latter are called "thin sandwich" initial value problems because they involve two nearby slices of spacetime.

In both classical and quantum dynamics, the "action" (Hamilton's principal function) is of fundamental importance. It is defined by

$$S = \int_{t_0}^{t_1} L\, dt\,. \tag{3}$$

One may express L as $L(q, \dot{q})$ or as $L = p_i\dot{q}^i - H(q, p)$. In classical mechanics, the system motion is determined, with *either* form of L, by finding the path $q(t)$ or $(q(t), p(t))$ that makes S an extremum for given fixed end points $q^i(t_0)$ and $q^i(t_1)$. In the sum-over-paths approach to quantum theory, S (divided by $\hbar$) becomes the quantum mechanical phase associated with every kinematically possible path from a fixed $q^i(t_0)$ to a fixed $q^i(t_1)$. With either form of L, one must specify q^i at two finitely separated times. In classical dynamics, or in the classical limit of the quantum dynamics, one thus has a "long-time" evolution problem from t_0 to t_1. As $(t_1 - t_0)$ becomes small, one recovers a thin-sandwich *initial value* problem. One must know how to specify $q^i(t_0)$ and $q^i(t_0+\delta t)$ in a consistent manner. It appears then, that thin sandwich problems are an essential feature of dynamics based directly on the action. If, as in general relativity or in gauge theories, there are conditions limiting the choices of initial data ("constraints"), then the thin sandwich problem (as well as the one-time initial value problem) becomes nontrivial. In general relativity there is the added complication of defining "time" in a suitable way.

In general relativity, then, the easier of these approaches is to pose and solve the initial value problem on one slice and to follow the evolution using the first order equations of motion. Because the main purpose of this article is to describe some of the recent progress in constructing solutions of the Einstein equations, we shall concentrate on the one-slice initial value problem.

We emphasize, however, that the scope of initial value problems is far wider than merely being "step one" of the integration of equations of motion. We therefore also regard our results on one-time initial value problems as being possibly important guides for an eventual resolution of the thin sandwich problem.

2. A FLAT SPACETIME INITIAL VALUE PROBLEM: MAXWELL'S THEORY

We next consider a familiar example of a relativistic field theory with constraints on the initial data: Maxwell's theory in Minkowski spacetime. This example will be useful in establishing a paradigm for the analysis of more complicated initial value problems, such as that of general relativity.

In spacetime, an "initial time" means a three-dimensional spacelike hypersurface Σ_0. The choice of such a slice is essentially arbitrary as far as the physics resulting from the solution of the associated Cauchy problem is concerned. However, in practice, it is useful to have a prescription that simplifies the description of the initial value problem (and the ensuing evolution of the initial data) as much as possible. The prescription will be expected to lean heavily on the geometry of the underlying spacetime.

Such a geometrical prescription for what are in essence the kinematical features of the Cauchy problem is, of course, readily available in Minkowski spacetime. Let the flat spacetime metric be denoted by $f_{\mu\nu}$, with $f_{\mu\nu} = \eta_{\mu\nu}$ in rectangular coordinates. Then a preferred time is defined by finding a timelike Killing vector field t^μ, with $t^\mu t_\mu = -1$ as a normalizing condition. That is, we solve Killing's equations

$$\pounds_t f_{\mu\nu} = \nabla_\mu t_\nu + \nabla_\nu t_\mu = 0\,, \tag{4}$$

where $\pounds$ denotes the Lie derivative and ∇ denotes the (torsionless) covariant derivative compatible with $f_{\mu\nu}$. The solution we seek satisfies also

$$t_{[\mu}\nabla_\nu t_{\alpha]} = 0 \tag{5}$$

and is therefore orthogonal to a family of spacelike slices Σ. Because $t^\mu t_\mu = -1$, we see that $t^\mu = n^\mu$ is the unit normal of these slices. We denote the vector field $t^\mu(\partial/\partial x^\mu)$ by $\partial/\partial t$ and adapt the time coordinate by setting $x^0 = t$. Then the family of slices Σ to which t^μ is orthogonal can be labeled by the values of t; we denote this by writing Σ_t. Now the Minkowski spacetime line element has the form ($c = 1$)

$$ds^2 = -dt^2 + f_{ij}(x^k)dx^i dx^j \,, \tag{6}$$

where f_{ij} is the flat three-metric of the slices Σ_t and $i, j, k, \ldots = 1, 2, 3$.

Of course, the above is just a geometrical description of choosing the usual Minkowski time coordinate. Note that this process involved finding a timelike translational Killing vector $t^\mu(\partial/\partial x^\mu) = \partial/\partial t$, labeling spacelike slices by $x^0 = t$, and choosing the slices to be orthogonal to the integral curves of $\partial/\partial t$. The preferred slices of Minkowski spacetime thus result from special symmetry conditions. There is a three-parameter set of such "foliations" of flat spacetime by flat spacelike hyperplanes obtained from the action of the Lorentz boosts on a given slicing.

Preferred slices that result from special symmetries do not in general exist in curved spacetimes. However, there is a simple geometrical variational principle that enables us to pick out the spacelike hyperplanes in Minkowski spacetime and that also carries over to curved spacetimes, where it enables us to pick out spacelike hypersurfaces with some of the same simplifying properties. One seeks spacelike hypersurfaces on which the volume of any bounded region is stationary with respect to any small deformation that vanishes on the boundary. Because the spacetime signature is $(-+++)$, these hypersurfaces turn out to have maximal volume and are therefore called maximal slices. The resulting differential equation states that $\nabla_\mu n^\mu = 0$, where n^μ is the unit timelike normal field. This condition is equivalent to the vanishing of the mean (extrinsic) curvature "trK" ($trK = -\nabla_\mu n^\mu$). However, it is known (Calabi, 1968; Cheng and Yau, 1976) that the only complete, smooth, maximal, spacelike hypersurfaces in Minkowski spacetime are the standard $t =$ constant hyperplanes. Thus, one can replace a stringent set of symmetry conditions by a more widely applicable geometrical principle whose satisfaction does not require the existence of symmetries of the spacetime geometry.

On a $t =$ constant hyperplane, we write the Maxwell equations in terms of the four-potential A_μ split into the scalar potential φ and the three-vector potential $A^j = f^{ji}A_i$:

$$A_\mu = (-\varphi, A_i) \,. \tag{7}$$

In addition, we introduce as an independent variable the electric field $E^j = f^{ji}E_i$. The electric field is subject to the constraint

$$D^iE_i = 4\pi\rho \,, \tag{8}$$

where D_i denotes the spatial covariant derivative associated with f_{jk} (i.e., $D_i f_{jk} = 0$).

Exercise: Write the action for Maxwell's theory in terms of A_μ and E^i, and show that the constraint results from the invariance of the action with respect to gauge transformations $A_\mu \to A_\mu + \partial_\mu \chi$, where χ is an arbitrary function.

The charge density ρ and the charge current density $j^j = f^{ji} j_i$ are the hyperplane-orthogonal and hyperplane-tangential pieces of the four-current

$$J_\mu = (-\rho, j_i) .$$

The initial value problem is to give $(\varphi, A_i, E_i, \rho, j_i)$ at $t = t_0$, subject to the constraint (8).

The initial value problem can be solved by decomposing the electric field on the initial slice into longitudinal and transverse (divergence-free) parts:

$$E_i = E_i^T + E_i^L , \quad E_i^L \equiv D_i U , \tag{9a}$$

$$D^i E_i^L = D^i D_i U \equiv \Delta U , \quad D^i E_i^T = 0 . \tag{9b}$$

If we regard E_i in (9*a*) as a one-form, we say the splitting is to find the "exact" ($D_i U$) and "co-exact" (E_i^T) parts of E_i. The only other kind of one-form is called "harmonic" (vanishing divergence *and* curl). There are no regular harmonic one-forms on $\mathbf{R}^3$ that vanish at infinity. See Misner and Wheeler (1957) for an elementary discussion.

The initial value equation is now, apart from an irrelevant sign, the usual Poisson equation

$$\Delta U = 4\pi\rho . \tag{10}$$

We notice that the initial value equation, as expected, is of elliptic type. It has, of course, nothing to do with any physical propagation of signals because it holds at a fixed time.

We wish to solve (10) such that $U \to 0$ for large spacelike distances from some origin (i.e., as $r \to \infty$). This will be possible provided that ρ goes to zero sufficiently rapidly. There is no problem in achieving this result in electrodynamics; one can usually take ρ with compact support. However, in the analogous problem in general relativity, where the field acts as its own source, the fall-off of sources is more delicate. Therefore, we note that $\rho = (r^{-(3+\varepsilon)})$, $\varepsilon > 0$, suffices. In such cases we can write

$$U = 4\pi\Delta^{-1}(\rho) , \tag{11}$$

and the initial value equation at $t = t_0$ is satisfied by

$$E_i = E_i^T + 4\pi D_i[\Delta^{-1}(\rho)] . \tag{12}$$

Here E_i^T is any divergence-free electric field; therefore, E_i^T represents the freely specifiable part of E_i at $t = t_0$.

Given any electric field satisfying (12) on the initial slice, we may proceed to the evolution problem. The evolution is determined by the first order equations of motion of the field

$$\partial_t A_i = -E_i - D_i \varphi\,, \tag{13}$$

$$\partial_t E_i = R_i - 4\pi j_i\,, \tag{14}$$

where R_i contains the second spatial derivatives of A_i:

$$R_i \equiv D^j D_j A_i - D_i D^j A_j\,. \tag{15}$$

Note that $D^i R_i = 0$. We must also require the local conservation of charge (source evolution equation)

$$\nabla^\mu J_\mu = 0 = \frac{\partial \rho}{\partial t} + D^i j_i\,. \tag{16}$$

It is of the first importance to see that the initial value constraint (8) need only be imposed once, that is, at $t = t_0$. This property is necessary for any true initial value constraint of a Cauchy problem. Otherwise, one could not distinguish between initial value equations and evolution equations. We state this result as an exercise.

Exercise: Show from the field and source evolution equations that (8) holds for any $t > t_0$ if it holds at $t = t_0$.

We also observe that there is no evolution equation for φ, that is, no term $\partial\varphi/\partial t$ in (13) and (14). This underdetermination of the evolution of (φ, A_i, E_i) is another consequence of the gauge freedom inherent in Maxwell's theory. In effect, the fact that there is an arbitrary choice of a gauge function $\chi(t, x)$ means that one of the potentials (φ, A_i) has its evolution determined by the choice of χ. (Of course, the electric field E_i and the magnetic field $B_i = 1/2\ \varepsilon_{ijk} D^j A^k$ are gauge invariant and evolve deterministically.) The evolution of φ can only be determined through the fixing "by hand" of gauge conditions on the potentials. Two familiar choices are the Coulomb gauge $D^j A_j = 0$ and the Lorentz gauge $\nabla^\mu A_\mu = \partial\varphi/\partial t + D^j A_j = 0$. (For a discussion of these gauges in the present context, see Smarr and York, 1978*a*.) In either case, one may combine Maxwell's equations (8), (13), (14), and the gauge condition to achieve second order "reduced" field equations of the form $\Box F = H$, where $F = A_i$ and $H = -4\pi j_i^T$ $(D^i j_i^T = 0)$ in Coulomb gauge and $F = A_\mu$, $H = -4\pi J_\mu$ in Lorentz gauge $(\Box \equiv -\partial_t \partial_t + \Delta)$.

Because F obeys an (inhomogeneous) wave equation, we see that the evolution equations are hyperbolic, in contrast to the elliptic constraint equation. Thus, the equations of motion refer to propagation of physical effects. The fundamental property of a ("normal") hyperbolic equation is (for a fixed source) that given F and its first normal derivative (here, $\partial_t F$) in a region B of the slice $t = t_0$, there is a domain of spacetime in which the solution F of $\Box F = H$ is unique. The quantities $(F, \partial_t F)$ on $t = t_0$ are called the Cauchy

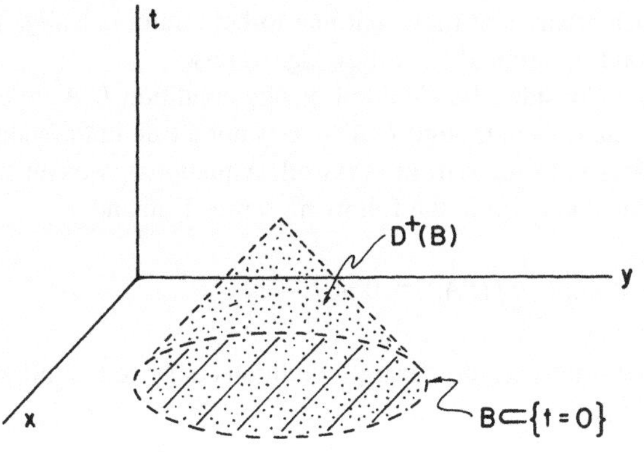

Figure 7.1. The "teepee" (and its interior) whose base coincides with the set B (here chosen to be a disk with a circular boundary on $t = 0$, i.e., a solid sphere in three-dimensional space) is the future domain of dependence of B, denoted $D^+(B)$.

data of the problem. For $t > t_0$ ($t < t_0$) the domain of spacetime in which the solution is unique is called the future (past) domain of dependence of the Cauchy data in B. The future domain of dependence $D^+(B)$ consists of the collection of all points P to the future of $t = t_0$ such that, if we construct the past light cone of any such P, including the interior of this cone, then the intersection with the slice $t = t_0$ is a set of events that is a subset of B. (See Figure 7.1.)

Exercise: Show that another way of stating the above uniqueness property is that the only solution of the homogeneous equation $\Box F = 0$ that has zero Cauchy data in $F(F = \partial_t F = 0$ in $B)$ is $F = 0$ in the domain of dependence of B. See Duff (1959).

The above discussion suggests several questions about any initial value problem:

Q1. On a spacelike hypersurface, what are the initial data and equations of constraint?

Q2. If the constraints hold on an initial slice, are they automatically satisfied at all future (or past) times by virtue solely of the evolution and source conservation equations?

Those initial conditions that continue to hold in the future by virtue of the field equations and conservation equations are the true initial value equations. Any remaining constraints are then effectively gauge conditions. These must be enforced "by hand" for all time during the evolution. Alternatively, they can *sometimes* be reformulated in terms of *initial* conditions and used in conjunction with special "reduced" or gauge-dependent forms of the equations of motion in such a way that they continue to hold automatically. This point is best understood in terms of the following exercise.

Exercise: Consider the Coulomb gauge condition $D_iA^i = 0$ imposed at $t = t_0$. Show that, even though $D_iA^i = 0$ is not a true initial value constraint (because it does not follow from Maxwell's equations), we can still treat it as an initial value condition in the following sense. Demand

$$D^iA_i = 0 \textit{ and } \frac{\partial}{\partial t}(D^iA_i) = 0 \text{ at } t = t_0 .$$

These two conditions are the "gauge condition converted to initial data conditions." Note that

$$\frac{\partial}{\partial t}(D^iA_i) \equiv D^i\left(\frac{\partial}{\partial t}A_i\right) .$$

Given the above, the constraint equation is

$$\Delta\varphi = -4\pi\rho \text{ at } t = t_0 .$$

Take as the "reduced" form of Maxwell's equations

$$\Box A_i = -4\pi j_i^T ,$$

where j_i^T is the divergence-free part of the source current. Show (a) that the above formulation is consistent with Maxwell's theory, and (b) that $D^iA_i = 0$ will hold for all time in this formulation.

Q3. How is the initial spacelike slice to be chosen?

Q4. How do we actually solve the constraint equations? Alternatively, which parts of the initial data can we specify freely such that the remaining parts are uniquely determined?

In addition, one may ask the following question about the evolution equations:

Q5. What are appropriate gauge conditions that render the evolution of the field variables ("potentials") unique in the domain of dependence?

Such gauge conditions will always be needed in any Cauchy problem in which there are initial value constraints among the field equations. No actual physical effect should depend on the choice of gauge. However, an inappropriate choice could cause the inadvertent exclusion of certain solutions or could make the solution of the evolution equations very difficult in practice (e.g., by numerical methods).

In the following we shall use the above questions to guide us in an overview of the initial value problem of general relativity. Little will be said about gauge and evolution problems, because these topics have received detailed attention in readily accessible recent works (York, 1979*a*). For similar reasons, few of the mathematical details of the initial value problem will be given. Instead, we shall describe several new solutions of the initial value equations as examples.

3. THE INITIAL VALUE PROBLEM FOR GRAVITY

In general relativity, one does not have a given spacetime in which to solve the field equations. One must construct the spacetime itself. Thus one must make some a priori assumptions about the smoothness, global topology, asymptotic conditions, etc., of the spacetime that is to be constructed. In particular, because we construct spacetime by solving a Cauchy problem, every spacetime that we consider must be "globally hyperbolic" with topology $\Sigma \times \mathbf{R}$, where Σ is a spacelike Cauchy data slice. We may take "globally hyperbolic" to mean the nonexistence of any closed or "almost closed" timelike curves. (See, for example, Hawking and Ellis, 1973.) The requirement of global hyperbolicity is in fact necessary in analyzing *any* relativistic field theory in terms of its Cauchy problem.

We proceed by finding suitable initial data on a spacelike slice Σ and calculating its evolution. It is known that there exists a maximal ("largest") domain of dependence for the initial data (Choquet-Bruhat and Geroch, 1969). However, at present no one knows an explicit construction of the maximal domain of dependence; it is only known to exist through an application of Zorn's lemma (axiom of choice). Nevertheless, excellent progress on explicit constructions has been made in recent studies (see, for example, Eardley and Smarr, 1979). We shall not worry about this problem in the present work.

We shall now take up questions *Q1–Q5* in order.

Q1. Initial Data and Constraints

The initial data of the gravitational field on a spacelike slice Σ are the positive definite metric g_{ij} of Σ and the second fundamental tensor (extrinsic curvature) $K_{ij} = K_{ji}$ of Σ. If the timelike unit normal of Σ is n^μ $(n^\mu n_\mu = -1)$, then the tensor $\mathbf{K}$ can be defined by

$$-2\mathbf{K} = \pounds_n \mathbf{g}\,, \tag{17}$$

where $\pounds$ denotes the Lie derivative. Thus, $\mathbf{K}(\Sigma)$ is essentially the "velocity" of $\mathbf{g}$ in the unit normal direction. This velocity is independent of the continuation of n^μ away from the given slice.

It is convenient to distinguish between the trace and traceless parts of K_{ij} by defining

$$E_{ij} \equiv K_{ij} - \frac{1}{3} g_{ij} trK\,, \tag{18}$$

$$trK \equiv g^{ij}K_{ij}\,; \tag{19}$$

thus, $K_{ij} = E_{ij} + 1/3\, g_{ij}trK$. Then we have that the mean curvature of Σ is $trK = -\nabla_\mu n^\mu = -$ (expansion of the vector field n^μ) $= +$ (convergence of n^μ) and that $\mathbf{E} = -$ (shear of n^μ).

Exercise: Verify the preceding interpretations of trK and E_{ij}. See Misner, Thorne, and Wheeler (1973), p. 517. This reference will be denoted by "MTW" below.

The initial data for the matter fields on Σ, if we disregard any independent initial value for the source fields per se, are the energy density ρ and the momentum density j^i. If we define the operator of projection onto Σ, $\perp^\mu_\nu = \delta_\nu^{\ \mu} + n^\mu n_\nu$, then $\rho = T^{\mu\nu}n_\mu n_\nu$ and $j^\mu = -\perp^\mu_\alpha T^{\alpha\nu}n_\nu$ $(j^\mu n_\mu = 0)$, where $T^{\mu\nu}$ is the stress-energy tensor. Taking $G = c = 1$ and using MTW conventions, we find that the constraint equations, when written entirely in terms of quantities defined on the *single* hypersurface Σ, are given by

$$D_jE^{ij} = \frac{2}{3}\nabla^i trK + 8\pi j^i\,, \tag{20}$$

$$R - E_{ij}E^{ij} + \frac{2}{3}(trK)^2 = 16\pi\rho\,, \tag{21}$$

where R is the scalar curvature of g_{ij} and D_j is the associated (spatial) covariant derivative operator. Geometrically, the constraints are simply the Gauss-Codazzi equations for the imbedding of Σ in a spacetime satisfying the Einstein equations $G_{\mu\nu} = 8\pi T_{\mu\nu}$. Formally, they arise from the spacetime coordinate (gauge) freedom of general relativity, as can be seen in the Arnowitt-Deser-Misner (1962) action principle.

Exercise: Deduce the Gauss-Codazzi equations

$$\perp {}^{(4)}R_{\mu\nu\alpha\beta} = R_{\mu\nu\alpha\beta} + \varepsilon(K_{\mu\beta}K_{\nu\alpha} - K_{\mu\alpha}K_{\nu\beta})\,,$$

$$\perp {}^{(4)}R_{\mu\nu\alpha\beta}n^\beta = D_\nu K_{\mu\alpha} - D_\mu K_{\nu\alpha}\,,$$

for the imbedding of Σ in a spacetime of signature $(\varepsilon, +, +, +)$ where $\varepsilon = \pm 1$. (The symbol $\perp$ denotes the projection of every free index in the tensor standing to the right.) Show that appropriate contractions of these equations yield the constraints if we set $\varepsilon = -1$ and use Einstein's equations (York, 1979*a*).

Q2. Preservation of Constraints

We can write Einstein's equations as $H^{\mu\nu} \equiv G^{\mu\nu} - 8\pi T^{\mu\nu} = 0$. In terms of $H^{\mu\nu}$, the constraints are given by $H^{\mu\nu}n_\mu n_\nu = 0$ and $\perp^\alpha{}_\mu H^{\mu\nu}n_\nu = 0$ or $H^{00} = 0$ and $H^{0i} = 0$. From the twice-contracted Bianchi identity of the spacetime Riemann curvature tensor, we know that $\nabla_\nu G^{\mu\nu} \equiv 0$. If we assume "matter conservation," we have $\nabla_\nu T^{\mu\nu} = 0$; thus $\nabla_\nu H^{\mu\nu} = 0$. These equations can be written as

$$H^{00}{}_{,0} = -(H^{0i}{}_{,i} + H^{0\alpha}\Gamma^{\nu}_{\alpha\nu} + H^{\nu\alpha}\Gamma^{0}_{\nu\alpha}) ,$$

$$H^{0i}{}_{,0} = -(H^{ij}{}_{,j} + H^{i\alpha}\Gamma^{\nu}_{\alpha\nu} + H^{\nu\alpha}\Gamma^{i}_{\nu\alpha}) ,$$

where Γ denotes the spacetime affine connection. It now follows from $\nabla_\nu T^{\mu\nu} = 0$ and the equations of motion $H^{ij} = 0$ that $H^{00} = H^{0i} = 0$ for all time if $H^{00} = H^{0i} = 0$ on an initial slice (here denoted by $t = x^0 =$ constant).

Q3. Choice of Initial Slice

This topic raises some of the most important and difficult questions in general relativity because it has to do with the nature of "time" in a theory that would seem to permit a great deal of arbitrariness in this concept. It is possible to take a "purist" point of view and say that the question is inherently difficult precisely because it is in fact unnatural. One can say that general relativity is a theory of spacetime, not of space and time, and in spacetime . . . nothing happens! It simply *is*. (Following Vonnegut, 1959, we can call this the "Tralfamadorian point of view.")

We can answer this objection in two ways. First, dynamical descriptions of all other successful physical theories, both in the classical and quantum domains, have proven to be very fruitful. In dynamics, one needs a useful notion of time. Therefore, in order to have a unified understanding of gravitational and other interactions, we must address the subject of time in general relativity. Second, in practice, one of the best workable methods of solving Einstein's equations in relatively complex, physically interesting situations is to integrate them, numerically, as a Cauchy problem. Therefore, we must find good ways to construct stacks of spacelike hypersurfaces. This means we must, at least implicitly, define a time variable.

We can divide this topic into three parts, of which we shall attend principally to the first: (1) initial construction of a slice, (2) embedding problems, and (3) foliation problems.

(1) In a construction problem, we simply give the value on the initial slice of some quantity that is very sensitive to the embedding of the slice in the spacetime to be constructed. The quantity that we use is the mean (extrinsic) curvature *trK*; it is in fact the simplest scalar that we can form from g_{ij} and K_{ij}. By construction, the spacetime will possess a slice with the given value of *trK*. We can get an idea of the geometric reason for using *trK* by considering the following fact: any deformation $\zeta(x)n^\mu$ of a spacelike slice with unit normal n^μ, where $\zeta(x)$ is nonnegative, not identically zero, and has compact support, necessarily changes the value of *trK* at least somewhere in the support of $\zeta(x)$. (Here we assume that the spacetime obeys the "strong

energy condition"; see *exercise* below.) From a physical point of view, the usefulness of *trK* in the present role can be traced to the idea that "gravity is always attractive" or "freely falling test particles tend to converge."

In practice the preferred choices of *trK* are, first, $trK = 0$ (in asymptotically flat spacetimes or in closed universes at the moment of maximum expansion), or second, $trK =$ const. $\neq 0$ (in open or closed universes or in asymptotically flat spacetimes, where such slices are spacelike but asymptotically null). These are, geometrically, the most natural choices, and in practice these choices simplify solution of the constraints, as we shall see.

(2) The embedding problem is, for a *given* spacetime, to find a spacelike slice with a prescribed value of *trK* (if such a slice exists). Such problems are quite difficult and only partial results are known. Here are a few: First, in any spacetime sufficiently close (in a suitable norm) to Minkowski spacetime, there exist maximal ($trK = 0$) spacelike slices. Second, Schwarzschild and Kerr black holes, and sufficiently "close" spacetimes, have maximal spacelike slices. Third, every stationary, axisymmetric, and asymptotically flat spacetime that is topologically $\mathbf{R}^4$ has maximal spacelike slices. Fourth, every homogeneous cosmological model (as well as "nearby" spacetimes) has spacelike slices of constant mean curvature ($trK =$ constant). The reason for studying the embedding problem is to determine whether or not any globally hyperbolic solution of the Einstein equations will be excluded if we restrict the value of *trK* on an initial slice (e.g., does every closed universe possess a constant mean curvature slice?).

(3) The foliation problem is concerned with the evolution of initial data and asks whether families of slices with a prescribed *evolution* of *trK*, given the initial such slice, exist. In fact, one knows that families of $trK = 0$ or $trK =$ constant slices always exist, given the first one, but much more needs to be discovered about the global (in time) behavior of such slicings (e.g., Do they avoid singularities? Do they cover "enough" of the maximal domain of dependence to enable us to answer physically interesting questions?).

Exercise: In a spacetime that satisfies the "strong energy condition" ($T^{\mu\nu}n_\mu n_\nu + \perp^\mu{}_\nu T^\nu_\mu \geq 0$ for all timelike unit vectors n^μ), consider a small pointwise deformation $\zeta(x)n^\mu$ of a spacelike slice with unit normal n^μ. Assume $\zeta(x) \geq 0$, $\not\equiv 0$, and that it has compact support on the slice. Show that *trK* must change at least somewhere in the support of $\zeta(x)$. (Hint: Use the equation for $\partial/\partial t\,(trK)$ given below in *Q5*.)

Q4. Method of Solution of Constraints

We shall describe a general method of converting the four constraint equations (20) and (21) into four semilinear elliptic equations for four "potentials" (a scalar and a three-vector). These equations constitute a relativistic generalization of the Poisson equation for the Newtonian gravitational potential.

First, as was implicit in the foregoing discussion, we shall group the data on a slice Σ in the following manner: $(g_{ij}, E_{ij}, trK; \rho, j^i)$. The mean curvature trK will be assumed to be given and fixed in the following discussion.

One of the key ideas is to specify the physical data listed above only up to conformal equivalence, an idea that was originated by Lichnerowicz (1944). For the metric g_{ij}, this means that we write

$$g_{ij} = \psi^4 \hat{g}_{ij}, \tag{22}$$

where $\hat{g}_{ij}$ (the "trial" or "base" metric) is given and ψ is a strictly positive function to be determined. (The ψ is called a "conformal factor" and is our scalar potential.) The transformation (22) is called "conformal" for the reason described below.

Exercise: If V^i and W^i are arbitrary vectors at a point P, show that the angle between them is invariant with respect to conformal transformations. If T^{ij} is a symmetric tensor, show that the eigendirections of T^{ij} with respect to g_{ij} are the same as those of $(\psi^r T^{ij})$, $r \varepsilon \mathrm{R}$, $|r| < \infty$, with respect to $\hat{g}_{ij} = \psi^{-4} g_{ij}$.

The inverse of $\hat{g}_{ij}$ is defined by $\hat{g}^{ij}\hat{g}_{jk} = \delta^i_k$ and the (torsionless) covariant derivative associated with $\hat{g}_{ij}$ satisfies $\hat{D}_i \hat{g}_{jk} = 0$. Hence, we have the following *exercise*.

Exercise: Show that the coefficients of the affine connections of g_{ij} and $\hat{g}_{ij}$ are related by

$$\Gamma^i_{jk} = \hat{\Gamma}^i_{jk} + 2\psi^{-1}(\delta^i_k \psi_{,j} + \delta^i_j \psi_{,k} - \hat{g}^{im}\hat{g}_{jk}\psi_{,m}). \tag{23}$$

An important formula is the one relating the scalar curvature R of g_{ij} to that $(\hat{R})$ of $\hat{g}_{ij}$. One finds from (23) that

$$R = \hat{R}\psi^{-4} - 8\psi^{-5}\hat{\Delta}\psi, \tag{24}$$

$$\hat{\Delta} \equiv \hat{g}^{ij}\hat{D}_i \hat{D}_j. \tag{25}$$

All of the above formulas follow simply from the definition of a conformal transformation of the metric.

For the other data, we choose conformal transformation rules that preserve some key property. The traceless part E^{ij} of K^{ij} is transformed using

$$E^{ij} = \psi^{-10}\hat{E}^{ij}, \tag{26}$$

because the zero trace is preserved and, more important, because

$$D_j E^{ij} \equiv \psi^{-10}\hat{D}_j \hat{E}^{ij} \text{ for all } \psi > 0. \tag{27}$$

Note that indices on quantities with a circumflex (such as $\hat{E}^{ij}$) are raised and lowered with the base metric $\hat{g}_{ij}$ (therefore, $E_{ij} = \psi^{-2}\hat{E}_{ij}$).

Any traceless symmetric tensor such as $\hat{E}^{ij}$ can be split into a part $(\hat{E}^{ij}_*)$ with vanishing divergence and trace and a part derived by differentiating a "vector potential," W^i. (Note that we will take $W^i \equiv \hat{W}^i$ but that $W_i = g_{ij}W^j = \psi^4 \hat{W}_i$.) Therefore, we can write (York, 1973)

$$\hat{E}^{ij} = \hat{E}_*^{ij} + (lW)^{ij}, \tag{28}$$

$$(lW)^{ij} \equiv \hat{D}^i W^j + \hat{D}^j W^i - \frac{2}{3}\hat{g}^{ij}\hat{D}_k W^k. \tag{29}$$

Exercise: The conformal Killing equations are defined by $\pounds_V g_{ij} = \lambda g_{ij}$, where λ is a function. Show that V^i is a conformal Killing vector of $g_{ij} = \psi^4 \hat{g}_{ij}$, $\psi > 0$, if and only if $(lV)^{ij} = 0$. (Hint: show $(lV)^{ij} \equiv \psi^{-4}(lV)^{ij}$ for all $\psi > 0$.) Note that the splitting (28) and (29) works in any number of dimensions $n \geq 3$ (in [29] replace "3" by "n") and that the conformal properties are unaffected (use $4[n-2]^{-1}$ instead of 4 as the exponent of ψ).

From the form of the momentum constraints (20), and from (28) and (29), we see that the operator that enters the momentum constraints to determine the vector potential is

$$\hat{D}_j(lW)^{ij} \equiv (\hat{\Delta}_l W)^i = (\hat{\Delta} W)^i + \frac{1}{3}\hat{D}^i(\hat{D}_j W^j) + \hat{R}^i_j W^j, \tag{30}$$

where $(\hat{\Delta}W)^i \equiv \hat{D}^j \hat{D}_j W^i$. The "vector Laplacian" $\hat{\Delta}_l$ has a number of interesting properties that are described in the following *exercises*.

Exercise: Show that $\hat{\Delta}_l$ is formally self-adjoint, elliptic, and strongly elliptic. Construct the "Dirichlet" or "energy" integral for $\hat{\Delta}_l$, and show that it is strictly positive for any nontrivial W^i that is not a conformal Killing vector. (Hint: See Duff, 1959, for elliptic operators on scalars and note that $\hat{\Delta}_l$, like $\hat{\Delta}$, is in "divergence form," $\hat{\Delta}\psi = \hat{D}_j(\hat{D}^j\psi)$.)

Exercise: In a three-dimensional Riemannian space, consider all second order linear vector operators in the particular divergence form

$$D_j(D^i W^j + D^j W^i - r g^{ij} D_k W^k),$$

where r is any real constant. Show that the only choice of r that produces an operator that is *not* elliptic is $r = 2$. (The case $r = 2$ is relevant to the Baierlein, Sharp, and Wheeler, 1962, "thin sandwich conjecture," and the lack of ellipticity is one of the apparent problems of this conjecture.)

Consider compact spaces without boundary ("closed"). From the properties of $\hat{\Delta}_l$ established above, it follows that $\hat{\Delta}_l$ can be inverted uniquely in

$$(\hat{\Delta}_l V)^i = \hat{M}^i \tag{31}$$

provided $\hat{M}^i$ is orthogonal to any conformal Killing vector C^i of $\hat{g}_{ij}$ (if indeed there are any such C^i's). This is because the C^i's are the elements of the kernel of $\hat{\Delta}_l$. Thus one would need

$$\int \hat{g}_{ij}\hat{M}^i C^j \sqrt{g}\, d^3x = 0 \tag{32}$$

as an "integrability condition." See Choquet-Bruhat and York (1980).

Exercise: Show that the integrability condition (32) is identically satisfied if $\hat{M}^i$ has the form $\hat{M}^i = \hat{\nabla}_j \hat{T}^{ij}$, where $\hat{T}^{ij} = \hat{T}^{ji}$ and $\hat{g}_{ij}\hat{T}^{ij} = 0$.

We note in passing that there are no integrability conditions on (31) for asymptotically flat metrics on R^n (Cantor, 1979*b*; York, 1974).

The tensor $\hat{E}_*^{ij}$ in (28) can itself be constructed by freely specifying a symmetric trace-free tensor $\hat{T}^{ij}$. We use the decomposition and obtain

$$\begin{aligned} \hat{E}_*^{ij} &= \hat{T}^{ij} - (\hat{l}V)^{ij}, \\ (\hat{\Delta}_l V)^i &= \hat{\nabla}_j \hat{T}^{ij}. \end{aligned} \tag{33}$$

Combining (28) and (33) gives

$$\hat{E}^{ij} = \hat{T}^{ij} - (\hat{l}V)^{ij} + (\hat{l}W)^{ij} = \hat{T}^{ij} + (\hat{l}X)^{ij} \tag{34}$$

where $X^i \equiv W^i - V^i$. In (34), $\hat{T}^{ij}$ is the freely specified part of $\hat{E}^{ij}$. The physical E^{ij} will be obtained by setting $E^{ij} = \psi^{-10}\hat{E}^{ij}$ once we have found ψ.

Nothing remains to be done in reformulating the initial value equations except to discuss the sources ρ and j^i in (20) and (21). Their detailed treatment depends on whether they arise from a fluid, another field or fields, or are merely thought of as functions $\rho(x)$ and $j^i(x)$ on the slice. Various cases have been discussed in detail in the literature (Isenberg, Ō Murchadha, and York, 1976; Isenberg and Nestor, 1977). Here we shall assume that we specify $\hat{\rho}(x)$ and $\hat{j}^i(x)$, where

$$\rho(x) = \psi^{-8}\hat{\rho}(x), \quad j^i(x) = \psi^{-10}\hat{j}^i(x). \tag{35}$$

This procedure has the virtue of guaranteeing for all $\psi > 0$, that if

$$\hat{\rho} \geq +(\hat{g}_{ij}\hat{j}^i\hat{j}^j)^{1/2}, \tag{36}$$

then

$$\rho \geq +(g_{ij}\, j^i j^j)^{1/2}. \tag{37}$$

Hence, both the "weak energy condition" and the "dominance of energy condition" (Hawking and Ellis, 1973) can be built in a priori without knowledge of ψ.

Exercise: Verify the above assertions about the two energy conditions by using (35) and (36).

We are now in a position to write the general form of the initial value equations as a semilinear elliptic system determining a scalar potential and a vector potential, thereby constituting the full relativistic generalization of Poisson's equation. These equations are

$$(\hat{\Delta}_l X)^i = 8\pi\hat{j}^i - \hat{\nabla}_j \hat{T}^{ij} + \frac{2}{3}\psi^6\hat{\nabla}^i trK \tag{38}$$

$$-8\hat{\Delta}\psi = -\hat{R}\psi - \frac{2}{3}(trK)^2\psi^5 + [\hat{T}^{ij} + (\hat{l}X)^{ij}]^2\psi^{-7} + 16\pi\hat{\rho}\psi^{-3}. \tag{39}$$

Note that whenever we choose trK such that $\partial_i trK = 0$, then (38) is linear, second order, elliptic, and is entirely independent of ψ!

The freely specified quantities in these equations are $\hat{g}_{ij}$, $\hat{T}^{ij}$, trK, $\hat{\rho}$, and $\hat{j}^i$. The quantities to be determined are ψ and X^i. The physical initial data are then given by

$$g_{ij} = \psi^4 \hat{g}_{ij}, \tag{40}$$

$$K^{ij} = \psi^{-10}[\hat{T}^{ij} + (\hat{l}X)^{ij}] + \frac{1}{3}\psi^{-4}\hat{g}^{ik}trK, \tag{41}$$

$$\rho = \hat{\rho}\psi^{-8}, \quad j^i = \hat{j}^i\psi^{-10}. \tag{42}$$

The physical interpretation of these quantities on an asymptotically flat slice with $trK = 0$ is particularly illuminating. It may be stated as follows: $\hat{\rho}$ and $\hat{j}^i$ determine the energy and momentum densities of the matter fields. There is in (38) an effective momentum density of "gravitational waves" given by $-\nabla_j\hat{T}^{ij}$, that is, the field acts as its own source if it is nonlinear and self-interacting. The tensor $\hat{T}^{ij}$ itself, when squared, is an effective kinetic energy density for the gravitational waves, and $-\hat{R}$ is their effective potential energy. The $\mathcal{O}(r^{-1})$ part of ψ at large r determines the total energy of the complete system, "waves plus matter." The $\mathcal{O}(r^{-1})$ part of X^i determines the total linear momentum, and the total angular momentum can be easily extracted from the $\mathcal{O}(r^{-2})$ part of X^i. See York (1979*a*), Smarr and York (1978*a*), and Ō Murchadha and York (1974*a,b*) for more details.

We should note that (38) and (39) differ slightly from the forms of these equations given in York (1979*a*) and Choquet-Bruhat and York (1980). However, they are completely equivalent. Existence and uniqueness on closed slices are found in the latter reference. For asymptotically flat slices, the best results at the time of writing are in Cantor (1979*a*).

There is an important question that may arise when one looks at the "energy constraint" (39). That is, since all energy densities have the same physical dimensions, why are they all scaled differently in (39)? The answer for closed slices is the following: It is true that all energy densities have the same dimensions, but we have *chosen* a scaling that will *make* them all "fit" into a closed universe that satisfies Einstein's equations, which must have *zero* total energy. That is, (39) can "almost always" (Choquet-Bruhat and York, 1980) be solved for $\psi > 0$ unlike similar equations with different powers of ψ; and if the slice Σ is closed ($\partial\Sigma = 0$), we have

$$16\pi E = -8 \int_\Sigma \hat{\Delta}\psi\sqrt{\hat{g}}\ d^3x = 0$$

by Gauss's theorem. This is analogous to the fact that a closed universe can have no net electric charge, according to Maxwell's theory. Hence, the different powers of ψ are just the factors that adjust the energy densities of various types to add up to zero!

In the case of asymptotically flat slices, we have $E \neq 0$, but there is an analogous argument. We have in principle a choice of how to scale everything conformally, except the metric and its curvature. If there is no energy-momentum of matter ($\hat{\rho} = \hat{j}^i = 0$, i.e., vacuum) and no gravitational kinetic energy or momentum ($K_{ij} = 0$, a "moment of time symmetry"), then the entire problem reduces to the vacuum time-symmetric initial value equation $R = 0$ or

$$-8\hat{\Delta}\varphi = -\hat{R}\varphi\,. \tag{43}$$

Suppose (43) has an acceptable solution $\varphi > 0$, $\varphi \to 1$ as $r \to \infty$. Then, according to an important theorem of Cantor (1979*a*), assuming only that $trK = 0$, then the complete system (38) and (39) always has a unique solution $\psi > 0$, $\psi \to 1$ as $r \to \infty$ ($\psi \neq \varphi$). (Moreover, if (43) has no acceptable solution, then neither does (39). Cantor and Brill [in press] have shown the existence of metrics $\hat{g}^{ij}$ such that (43) cannot be solved. There is, in such cases, "too much gravitational wave potential energy," for example, see the Wheeler-Brill "closure as an eigenvalue problem" argument in Wheeler, 1963.) Hence, the particular scalings we have chosen automatically adjust all the energies and momenta, *except* for the gravitational potential energy $-\hat{R}$, which we assume to be acceptable, in such a way as to ensure that the complete problem has a solution. Moreover, the solution has nonnegative energy, as follows from the work of Brill (1959), Ō Murchadha and York (1974*b*), and Jang (1979). (The general proof of the full positive energy conjecture [York, 1980] has recently been given by Schoen and Yau, 1980. Note that *every* previous attempt at proving the full conjecture was, in one way or another, inadequate [York, 1980].)

Q5. Kinematical Conditions

The full form of the spacetime metric is given by

$$ds^2 = -(\alpha^2 - g^{ij}\beta_i\beta_j)dt^2 + 2\beta_i dx^i dt + g_{ij}dx^i dx^j\,, \tag{44}$$

where one uses the identity

$$K_{ij} = \frac{1}{2\alpha}\left[D_i\beta_j + D_j\beta_i - \frac{\partial g_{ij}}{\partial t}\right] \tag{45}$$

to relate (44) and the initial data (g_{ij}, K_{ij}). The kinematic or gauge problem is to give a useful specification of the time derivative

$$\frac{\partial}{\partial t} = \alpha n + \beta^i \frac{\partial}{\partial x^i}, \tag{46}$$

where α is the lapse function, n is the unit normal of t = constant, and β^i is the shift vector. See, for example, York (1979*b*) for a detailed discussion. We offer here only the barest sketch of the "gauge" problem.

Just as in electrodynamics, one may impose hyperbolic conditions analogous to the Lorentz gauge, and just as in electrodynamics, these conditions are automatically conserved by a special "reduced" form of the field equations (see, for example, Choquet-Bruhat and York, 1980). In general relativity these are the "harmonic" conditions

$$^{(4)}g^{\mu\nu(4)}\Gamma^{\alpha}_{\mu\nu} = 0 .$$

However, these are noncovariant even if a choice of time-slicing has been made. It is not known if they help in any way if we are constructing a spacetime starting on an initial data surface. See for more discussion Smarr and York (1978*a*, 1978*b*).

There are useful elliptic gauges in general relativity, somewhat analogous to the Coulomb gauge in electrodynamics (Smarr and York, 1978*a*). These may involve preserving the maximal slicing condition $trK = 0$ and using a "minimal distortion" shift vector or other possible choices. See Eppley (1979), Smarr (1979), Wilson (1979), and York (1979*b*). The elliptic gauges must be "enforced" repeatedly at each time step of the evolution. Unlike the case of the Coulomb gauge discussed previously in the present article, no way of imposing the elliptic gauges as "initial value conditions" that are automatically conserved by some special form of the Einstein equations is known at present. The solution of such a problem in an explicitly useful form would be a major step, but we do not even know if it is possible.

4. NUMERICAL METHODS

The equation (39) for the conformal factor ψ is the most general form that needs to be considered, unless, as is sometimes done, the matter term with $\hat{\rho}$ is modified slightly. It has four "source" terms for the Laplacian: (*a*) the scalar curvature $\hat{R}$, (*b*) the kinetic term $[\hat{T}^{ij} + (\hat{l}X)^{ij}] = \hat{E}_{ij}\hat{E}^{ij}$, (*c*) the trK term, and (*d*) the matter term $\hat{\rho}$. This equation decouples from the vector constraint (38) whenever trK = constant. (See Eardley and Smarr, 1979; Smarr and York, 1978*b*; Cantor et al., 1976; and Choquet-Bruhat, 1975, for discussions of the existence of these slices.) We shall assume here that the vector constraint has been solved analytically (as in the next section) or numerically.

The ψ constraint (39) has four limits in which all but one of the source terms vanish. Eppley (1977) solved this equation for time-symmetric Brill waves (Brill, 1959). In this case only the "curvature potential" $\hat{R}$ is present, and there are methods, like SOR (successive overrelaxation, Ames, 1977), that are applicable and prove to converge. Eppley (1979) avoided the nonlinear $(\hat{E}_{ij})^2$ term by rescaling $\hat{E}_{ij}$, but this recouples ψ to the vector constraint, a route we wish to avoid. P. Chrzanowski (personal communication, 1977) derived a matter limit ($\hat{R} = (\hat{E}_{ij})^2 = trK = 0$) solution to (39). His method, however, is applicable only to a specific choice of matter distribution.

Here we formulate a general numerical scheme for solving (39) that resembles the usual SOR method. The unknown ψ is defined at discrete points denoted (i, j, k) for the three coordinates, and $\hat{\Delta}$ is replaced by a finitely differenced operator $\hat{\Delta}_{FD}$. Equation (39) then becomes a set of nonlinear coupled algebraic equations for ψ_{ijk}:

$$\hat{\Delta}_{FD}\psi_{ijk} + S[\psi_{ijk}] = 0\,, \tag{47}$$

$$S[\psi] = -\hat{R}\psi + \frac{1}{8}\hat{E}^i_j\hat{E}_i^{\ j}\psi^{-7} - \frac{2}{3}(trK)^2\psi^5 + 2\kappa\hat{\rho}\psi^{-3}\,, \tag{48}$$

or

$$\psi_{ijk} + \frac{1}{a_{ijk}}\sum_{lmn}{}' a_{lmn}\psi_{lmn} + \frac{1}{a_{ijk}}S[\psi_{ijk}] = 0\,, \tag{49}$$

where Σ' does not include i, j, k and a_{lmn} are the coefficients of $\hat{\Delta}_{FD}$($\kappa = 16\pi$ in [48], if $G = c = 1$). There is no proof of existence or uniqueness of solutions of (49). Still, we follow the SOR method and attempt to solve it by successive overrelaxation. Given an approximation solution $\psi^{(n)}_{ijk}$, we derive the next $(n + 1)$ approximation by

$$\psi^{(n+1)}_{ijk} = \lambda\,\tilde{\psi}_{ijk} + (1 - \lambda)\psi^{(n)}_{ijk}\,, \tag{50}$$

with λ a free parameter (usually $1 \leq \lambda \leq 2$), and with $\tilde{\psi}_{ijk}$ defined by

$$F[\tilde{\psi}_{ijk}] = \tilde{\psi}_{ijk} + \frac{1}{a_{ijk}}S[\tilde{\psi}_{ijk}] + \frac{1}{a_{ijk}} \left[\sum_{lmn}^{(1)}{}' a_{lmn}\psi^{(n+1)}_{lmn} + \sum_{lmn}^{(2)}{}' a_{lmn}\psi^{(n)}_{lmn}\right]. \tag{51}$$

$\Sigma'^{(1)}$ is taken over values for which there is already an $(n + 1)$ approximation, and $\Sigma'^{(2)}$ is over the rest. (Neither sum includes i, j, k.) For nonlinear S, $\tilde{\psi}_{ijk}$ is calculated by Newton's method, that is, by using inner iterations with

$$\tilde{\psi} \rightarrow \tilde{\psi} - F[\tilde{\psi}]\left[1 + \frac{1}{a_{ijk}}\frac{\partial S}{\partial \tilde{\psi}}\right]^{-1} \tag{52}$$

until we achieve

$$F[\tilde{\psi}] < \varepsilon . \tag{53}$$

Note that

$$\frac{\partial S}{\partial \psi} = -\hat{R} - \frac{7}{8}\hat{E}^i_j \hat{E}_i^{\ j} \psi^{-8} - \frac{10}{3}(trK)^2 \psi^4 - 6\kappa\hat{\rho}\psi^{-4} . \tag{54}$$

Since $a_{ijk} < 0$ (always), $\psi > 0$, and for $\hat{R} > 0$, the denominator in equation (52) does not vanish. For a negative-bounded $\hat{R}$, a_{ijk} can be chosen large enough so that the denominator does not vanish. However, for $\hat{R} < 0$ it can happen that ψ does not remain positive.

Following Eppley (1977), we write $\hat{\Delta}_{FD}$ in a conservative form. However, we find a second order form is sufficient for an accurate solution. For a complete formulation of the problem, we must supply boundary conditions for ψ. The inner boundary (if any) and the inner boundary conditions depend on the specific problem. Asymptotic flatness determines the outer boundary condition, that is, one has

$$\psi = 1 + \frac{E}{2r} + \mathcal{O}(r^{-2}) . \tag{55}$$

The usual Dirichlet condition, $\psi = 1$ at the outer boundary, is inadequate for our problem. The boundary is at a finite radius, and $\psi = 1$ makes the gradient of ψ too steep. This means we do not obtain an accurate value of the total energy E, which is unacceptable! To formulate the outer boundary condition accurately at $r = R$, we use a Robin boundary condition (Duff, 1959). Thus, differentiate (55) with respect to r, and substitute the result into (55) to eliminate E. We find

$$\frac{\partial \psi}{\partial r}(R) + \frac{1}{R}(\psi - 1) = \mathcal{O}(R^{-3}) \tag{56}$$

In practice we take the right-hand side to be zero. This Robin condition replaces the usual inaccurate Dirichlet condition and enables us to achieve higher accuracy with smaller numerical grids. This condition is essential in the examples we solve in the next section.

One can see in Duff (1959) that Robin conditions give unique solutions, whenever they exist, to equations of the form $\mathcal{M}u = h$, where

$$\mathcal{M} = q^{ij}(x)\partial_i\partial_j + b^i(x)\partial_i + c(x) , \tag{57}$$

whenever $\mathcal{M}$ is elliptic and $c \leq 0$, just as do the more usual Dirichlet or Neumann boundary conditions. The general form of the Robin boundary condition is

$$\frac{\partial u}{\partial n} + f(x)u = l(x), \tag{58}$$

where $f(x) \geq 0$ on the boundary, whose outward pointed normal is n. Cantor has shown in unpublished notes from 1979 that one also obtains uniqueness for nonlinear elliptic equations such as those treated in the next section.

If the momentum constraint cannot be solved analytically, it can also be treated by finite differencing and SOR. It is elliptic and linear but of course is a coupled system of three equations. (Sometimes, however, the three equations can be uncoupled from each other.) If a numerical solution is needed, one is again forced to choose appropriate outer boundary conditions at $r = R$, R finite. Again, the fact that $X^i \to 0$ as $r \to \infty$ does not help. One turns to *vector Robin conditions*, which, as far as we know, have not appeared in the literature. They can be derived from the fact that the $\mathcal{O}(r^{-1})$ part of X^i is the linear momentum P^i. A treatment analogous to that for ψ (York, 1979*b*) gives the result at $r = R$:

$$\begin{aligned}(\hat{l}X)^{kj}e_j\left(\delta^i_k - \frac{1}{2}e^ie_k\right) + \frac{6}{7R}X^k\left(\delta^i_k - \frac{1}{8}e^ie_k\right) \\ = \mathcal{O}(R^{-3}),\end{aligned} \tag{59}$$

where e^i denotes the outward pointing unit normal. We would take the right-hand side to be zero in practice. It is not difficult to show that (59) guarantees that any solution X^i of an equation of the form

$$(\hat{\Delta}_l X)^i = V^i \tag{60}$$

is unique, as desired, if R is sufficiently large.

5. NEW INITIAL DATA FOR BLACK HOLES AND BLACK HOLE COLLISIONS

Recently Bowen and York (1980) have given an explicit construction of new initial data for a black hole or for two black holes that will undergo a non–head-on collision. Here we shall describe two different cases of data for single black holes. The method of solution uses the techniques described in the previous sections together with the method of "inversion through a sphere" or "Kelvin transformation." (See, for example, Jackson, 1962.)

One takes the initial slice to be maximal ($trK = 0$) and conformally flat ($g_{ij} = \psi^4\hat{g}_{ij}$, $\hat{g}_{ij} = f_{ij}$, f_{ij} = a flat three-metric). The initial slice, like the time-symmetric slice of Schwarzschild-Kruskal spacetime, has topology $S^2 \times$ R, with two asymptotically flat "ends." However, the new data will not

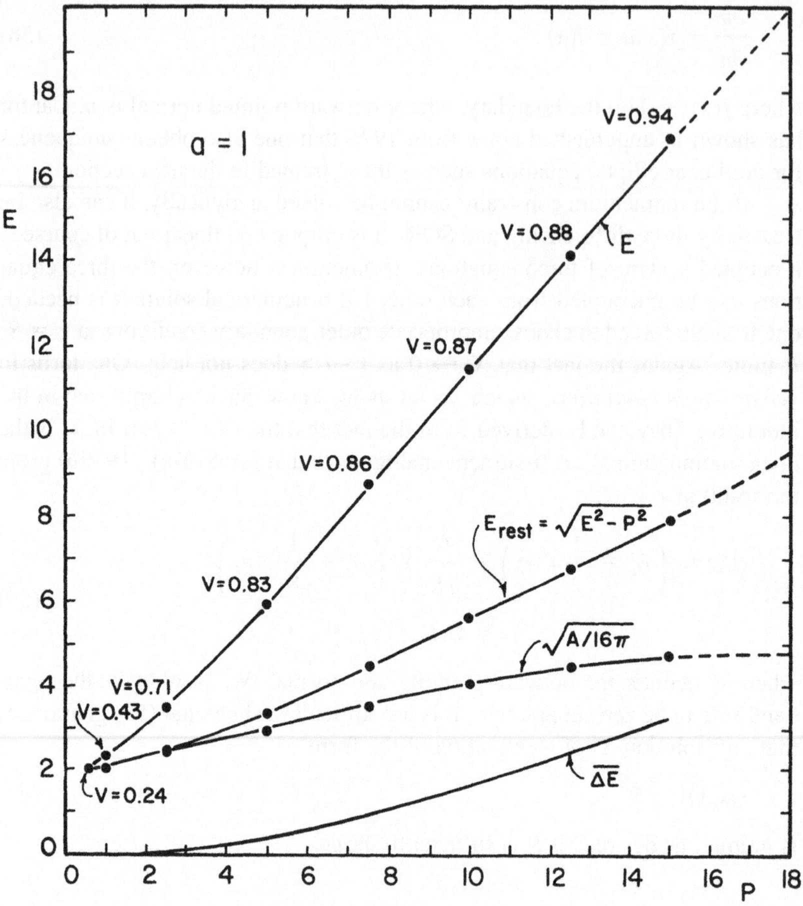

Figure 7.2. Energy of the boosted black hole. The energies E, $E_{rest} = (E^2 - P^2)^{1/2}$, M_{AH}; and $\overline{\Delta E} = E_{rest} - M_{AH}$ are plotted for values of $P = \mid P^i \mid$ ranging from $P = 0.5$ to $P = 15$ ($a = 1$). The velocity of the hole is defined by $V = PE^{-1}$. Note that $V \approx 0.94$ ($c = 1$) for $P = 15$.

generate a Schwarzschild-Kruskal spacetime, nor a Kerr spacetime, and will not be time-symmetric.

There is on the initial slice a minimal surface or "throat" located at $r = a$, where r is the usual flat-space spherical coordinate. However, this throat will not be spherically symmetric in the physical metric g_{ij}, although it will be a minimal surface. The "radius" a is a basic freely specifiable scale parameter of the construction ($a > 0$). Unlike the cases of Schwarzschild or Kerr black holes, or two black holes at a moment of time symmetry (Misner, 1960; Brill and Lindquist, 1963; Smarr, 1979), it is *not* possible by this technique to build in the mass parameter m in advance; one must find it as part of the solution.

Using inversion symmetry, conformal transformations, and the "vector Laplacian" $\hat{\Delta}_l$ introduced previously, one finds that data corresponding to a system with a total linear momentum P^i have

$$\hat{E}_{ij} = \frac{3}{2r^2}\,[P_i e_j + P_j e_i - (f_{ij} - e_i e_j)P^k e_k] + \frac{3a^2}{2r^4}\,[P_i e_j + P_j e_i + (f_{ij} - 5e_i e_j)P^k e_k]\,. \tag{61}$$

The constant vector P^i is freely specified (as is a) and turns out to be the total physical linear momentum. The second term in $\hat{E}_{ij}$ is needed to ensure that the physical data (g_{ij}, K_{ij}) will have $r = a$ as a minimal surface. The total energy E is found from the conformal factor ψ and is therefore dependent on P^i and a. (Actually, E is scaled by choosing a and $(P^i a^{-1})$.)

The conformal factor ψ is determined by (39), which in the present case is

$$\nabla^2\psi = -\frac{1}{8}\hat{E}_{ij}\hat{E}^{ij}\,\psi^{-7}\ (r \geq a)\,, \tag{62}$$

with boundary conditions

$$\frac{\partial\psi}{\partial r} + \frac{1}{2a}\psi = 0\,,\ \text{for } r = a\,, \tag{63}$$

$$\lim_{(r\to\infty)} \psi = 1\,. \tag{64}$$

The first ("inner") boundary condition is derived from an inversion symmetry requirement $\psi(r) = a/r\,\psi(a^2/r)$ and has the effect, together with the form of $\hat{E}_{ij}$, of ensuring an inversion-symmetric solution with $r = a$ as a minimal surface. In practice, in place of the second condition (64), we use a Robin condition, as described previously. Therefore, in the numerical solution for ψ, the region $r < a$ is eliminated and difficulties with the value to choose for ψ at large r (edge of the grid) are circumvented by the Robin condition. The latter gives an accurate value for E, which is essential for the physical interpretation of the results.

In the present case, it can be verified that the physical $K_{ij} = \psi^{-2}\hat{E}_{ij}$ will not satisfy $K_{ij}s^i s^j = 0$ at $r = a$, where s^i is the unit normal of $r = a$. Therefore, the minimal surface will *not* coincide with an apparent horizon. However, there *is* an apparent horizon slightly outside of the minimal surface, as is confirmed by a careful numerical solution of the apparent horizon equation, whose form is given by Bowen and York (1980). Because there is an apparent horizon, we see that our data correspond to a black hole, if we assume that no naked singularities arise in the evolution of this data (Hawking and Ellis, 1973). (The evolution has not been carried out at the time of writing.) That this new "boosted" black hole is not a boosted purely Schwarzschild black hole has been demonstrated by Bowen and York (1980). Its evolution will inevitably involve "real" gravitational radiation. We may describe it as a finitely perturbed boosted Schwarzschild black hole.

A plot of a typical ψ, even for quite large values of P (with a normalized to 1), does not clearly show the rather small but real deviations from spherical symmetry, so such a plot will not be given here. Also, the apparent horizon shows only small deviations from spherical symmetry. Hence, an embedding diagram for the solution closely resembles the familiar Einstein-Rosen bridge configuration.

To calculate $E \equiv E_\infty$, we use the formula (Brill, 1959; Ō Murchadha and York, 1974*b*)

$$E = -\frac{1}{2\pi}\oint D^i\psi d^2S_i, \tag{65}$$

where the integral is taken over the outer boundary. This energy contains contributions from the kinetic energy of motion, the mass of the hole, and the gravitational waves surrounding the hole. We define the rest mass as $E_{rest} = (E^2 - P^2)^{1/2}$. This includes the mass of the hole and gravitational radiation. See Figure 7.2. Note that we always have $E > P$ (no tachyons!) and that E_{rest} is therefore always real. This result is, of course, anticipated, but it was not imposed on the solutions during the calculations. It confirms the full positive energy conjecture (York, 1980; Schoen and Yau, 1980).

Another important quantity is a mass M_{AH} defined by

$$M_{AH} = \left(\frac{A_{AH}}{16\pi}\right)^{1/2}, \tag{66}$$

where A_{AH} is the area of the apparent horizon. Using the area theorem (Hawking and Ellis, 1973), we argue that M_{AH} is a lower limit on the irreducible mass of this black hole. Note that $E_{rest} > M_{AH}$ in Figure 7.2. The quantity

$$\overline{\Delta E} = E_{rest} - M_{AH} \tag{67}$$

gives an upper limit on the amount of gravitational wave energy in the system.

New initial data for a rotating black hole were constructed by very similar methods. In this case we have (Bowen and York, 1980)

$$\hat{E}_{ij} = \frac{3}{r^3}\,[\varepsilon_{kil}J^l e^k e_j + \varepsilon_{kjl}J^l e^k e_i]\,. \tag{68}$$

Note that a does not appear here. J^l is the constant, freely specified, total physical angular momentum of the system. The equation and boundary conditions on ψ are the same as those for the boosted hole, except that this new $\hat{E}_{ij}$ is used. The two-surface $r = a$ is again a nonspherical minimal surface on a conformally flat maximal slice of topology $\mathbf{R} \times S^2$. Bowen and York (1980) have shown that this data will not generate the Kerr solution, although it is a rotating black hole. There is no contradiction with the black hole uniqueness theorems because this data will not generate a stationary spacetime. Gravitational radiation is present.

However, because of the uniqueness theorems, and the perturbation analysis of Price (1972*a*,*b*), we expect that this rotating black hole will evolve asymptotically to a Kerr black hole, after the emission (or, possibly, absorption) of gravity waves. Therefore, a comparison with the Kerr solution is useful. (Our data are constructed on a "(t, φ)-symmetric" maximal slice [Hawking, 1973] and are compared to Kerr data on a (t, φ)-symmetric maximal slice.)

There are two natural energies to consider, $E = E_\infty$ and M_{AH}. In the Kerr solution, $M_{AH} = M_{IR}$, the irreducible mass, while E corresponds to the usual mass M of the Kerr solution. All of our solutions for the new hole satisfy $E > M_{AH}$. Clearly, however, the new E and M_{AH} do not satisfy the Christodolou (1970) formula relating M, J, and $M_{AH} = M_{IR}$ of Kerr. Therefore, we shall define two quantities analogous to the standard Kerr parameter, which is

$$\varepsilon_K = \frac{J}{M^2}\,, \tag{69}$$

where $\varepsilon_K = 1$ for an extreme Kerr hole.

The first one is

$$\varepsilon_l = \frac{J}{E^2}\,, \tag{70}$$

which gives a lower limit to the final angular momentum parameter of the new solution. An upper limit would be JM_{AH}^{-2}. However, a better approximation to the upper limit is

$$\varepsilon_u = J\left[M_{AH}^2 + \frac{J^2}{4M_{AH}^2}\right]^{-1}\,, \tag{71}$$

where we have applied the Christodolou formula with $M_{IR} \to M_{AH}$. Defining

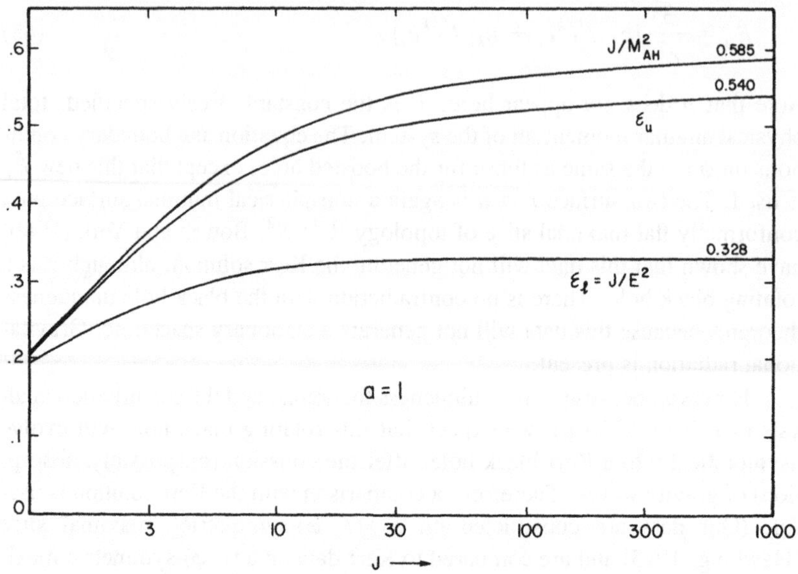

Figure 7.3. Angular momentum parameters as functions of $J(a = 1)$. The dimensionless angular momentum parameters "ε" that measure the rapidity of rotation for the new rotating black hole data are plotted against the total angular momentum J (with $a = 1$). The value $\varepsilon = 1$ would represent the maximally rotating hole in the case of a Kerr black hole.

$$M^2(M_{AH}, J) = M_{AH}^2 + \frac{J^2}{4M_{AH}^2}, \tag{72}$$

we find $M(M_{AH}, J) < E$. The difference

$$\overline{\Delta E} = E - M(M_{AH}, J) \tag{73}$$

is an upper limit to the amount of gravitational wave energy present. We are assuming in the present analysis that J = the total angular momentum of the system will be also the angular momentum of the final Kerr hole to which the new rotating hole is asymptotic. Thus, we are assuming that the hole will lose very little angular momentum as it radiates, and separate estimates not given here justify this completely.

As J increases ($a = 1$), ε_l goes asymptotically to 0.328. See Figure 7.3. It also appears that ε_u increases slowly to an asymptotic limit. For $J = 10^3$ we have $\varepsilon_u = 0.540$, and for $J = 10^4$ we have $\varepsilon = 0.550$. Hence, this family of solutions does not seem to have an extreme Kerr limit ($\varepsilon = 1$). We note

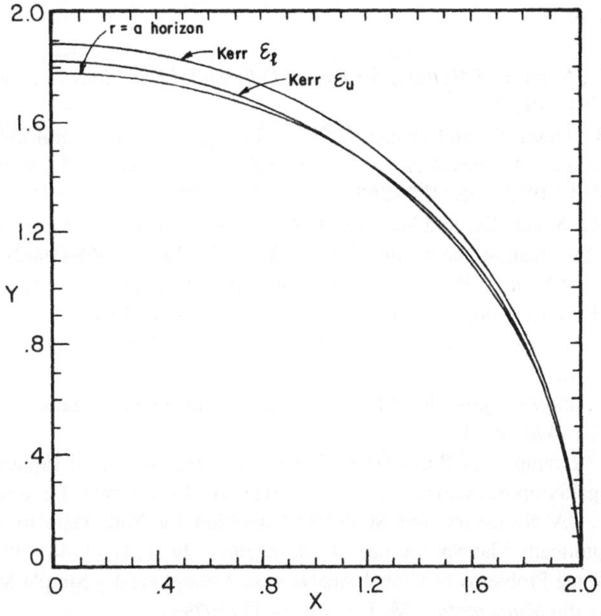

Figure 7.4. Embedding diagrams. Embedding of the $r = a$ horizon of the new black hole data and of Kerr horizons with corresponding ε_u and ε_ℓ (cf. Smarr, 1973).

that $\overline{\Delta E}$ is 25% of E when $J = 10^3$. So, conceivably, such a black hole could radiate a large amount of gravitational energy. However, a definite result awaits the evolution of this data, which has not yet been carried out.

In Figure 7.4, we compare the shape of the apparent horizon of the new hole to the Kerr geometry. We use $r = a$ for the new hole. (The minimal surface *is* the apparent horizon in this case.) We take two cases for the Kerr geometry. In one, we "normalize" Kerr by fixing its ε_K to be equal to the ε_l for the new hole; in the other we normalize by setting $\varepsilon_K = \varepsilon_u$ of the new black hole. The embeddings are in flat three-dimensional space and show the shapes of the equatorial planes (cf. Smarr, 1973). The new horizon of the new black hole is more flattened (more deformed) than the corresponding Kerr horizons, but does not reach the high deformations that are possible for rapidly rotating Kerr black holes.

In summary, these new black hole data appear to be interesting and significant and demonstrate the feasibility of finding new astrophysically interesting solutions of the Einstein equations using numerical techniques applied to the theory in its dynamical or "3 + 1" form.

REFERENCES

Ames, W. F., *Numerical Methods for Partial Differential Equations* (Academic Press, New York, 1977).

Arnowitt, R., Deser, S., and Misner, C. W., "The Dynamics of General Relativity," in *Gravitation—An Introduction to Current Research* (edited by L. Witten, Wiley, New York, 1962), pp. 227–265.

Baierlein, R., Sharp, D., and Wheeler, J. A., "Three-dimensional Geometry as Carrier of Information about Time," *Phys. Rev.* **126**, 1864–1865 (1962).

Bowen, J., and York, J. W., Jr., "Time Asymmetric Initial Data for Black Holes and Black Hole Collisions," *Phys. Rev. D* **21**, 2047–2056 (1980).

Brill, D. R., "On the Positive Definite Mass of the Bondi-Weber-Wheeler Time Symmetric Gravitational Waves," *Ann. Phys. (N. Y.)* **7**, 466–483 (1959).

Brill, D. R., and Lindquist, R., "Interaction Energy in Geometrostatics," *Phys. Rev.* **131**, 471–476 (1963).

Calabi, E., "Examples of Bernstein Problems for Some Non-linear Equations," Proceedings Symposia Global Analysis, University of California, Berkeley, 1968.

Cantor, M., "A Necessary and Sufficient Condition for York Data to Specify an Asymptotically Flat Spacetime," *J. Math. Phys.* **20**, 1741–1744 (1979*a*).

———, "Some Problems of Global Analysis on Asymptotically Simple Manifolds," *Compositio Mathematica* **38**, Fasc. 1, 3–35 (1979*b*).

Cantor, M., and Brill, D., "The Laplacian in Asymptotically Flat Manifolds and the Specification of Scalar Curvature," *Compositio Mathematica* (in press).

Cantor, M., Fischer, A., Marsden, J., Ō Murchadha, N., and York, J. W., Jr., "The Existence of Maximal Slices in Asymptotically Flat Spacetimes," *Commun. Math. Phys.* **49**, 189–190 (1976).

Cheng, S. Y., and Yau, S. T., "Maximal Space-like Hypersurfaces in the Lorentz-Minkowski Spaces," *Ann. of Math.* **104**, 407–419 (1976).

Choquet-Bruhat, Y., "Quelques propriétés des sous-variétés maximales d'une variété Lorentzienne," *C. R. Acad. Sci. A. (Paris)* **281**, 577–580 (1975).

Choquet-Bruhat, Y., and Geroch, R., "Global Aspects of the Cauchy Problem in General Relativity," *Commun. Math. Phys.* **14**, 329–335 (1969).

Choquet-Bruhat, Y., and York, J. W., Jr., "The Cauchy Problem," in *General Relativity and Gravitation* (edited by A. Held, Plenum Press, New York, 1980), pp. 99–172.

Christodolou, D., "Reversible and Irreversible Transformations in Black-hole Physics," *Phys. Rev. Letters* **25**, 1596–1597 (1970).

Duff, G. F. D., *Partial Differential Equations* (Univ. of Toronto Press, Toronto, 1959).

Eardley, D., and Smarr, L., "Time Functions in Numerical Relativity," *Phys. Rev. D* **19**, 2239–2259 (1979).

Eppley, K., "Evolution of Time-symmetric Gravitational Waves: Initial Data and Apparent Horizons," *Phys. Rev. D* **16**, 1609–1613 (1977).

———, "Pure Gravitational Waves," in *Sources of Gravitational Radiation* (edited by L. Smarr, Cambridge Univ. Press, Cambridge, 1979), pp. 275–291.

Hawking, S. W., "The Event Horizon," in *Black Holes* (edited by C. DeWitt and B. DeWitt, Gordon and Breach, New York, 1973), pp. 1–55.

Hawking, S. W., and Ellis, G. F. R., *The Large Scale Structure of Spacetime* (Cambridge Univ. Press, Cambridge, 1973).

Isenberg, J., Ō Murchadha, N., and York, J., "The Initial Value Problem of General Relativity," *Phys. Rev. D* **13**, 1532–1537 (1976).

Isenberg, J., and Nestor, J., "Extension of the York Field Decomposition to General Gravitationally Coupled Fields," *Ann. of Phys.* **108**, 368–386 (1977).

Jackson, J. D., *Classical Electrodynamics* (Wiley, New York, 1962).

Jang, P. S., "On Positivity of Mass for Black Hole Spacetimes," *Commun. Math. Phys.* **69**, 257–268 (1979).

Lichnerowicz, A., "L'intégration des Equations de la Gravitation Relativiste et le Problème des *n* corps," *J. Math. Pures Appl.* **23**, 37–63 (1944).

Misner, C. W., "Wormhole Initial Conditions," *Phys. Rev.* **118**, 1110–1112 (1960).

Misner, C. W., Thorne, K., and Wheeler, J. A., *Gravitation* (Freeman and Co., San Francisco, 1973).

Misner, C. W., and Wheeler, J. A., "Classical Physics as Geometry: Gravitation, Electromagnetism, Unquantized Charge, and Mass as Properties of Curved Empty Space," *Ann. Phys. (N. Y.)* **2**, 525–603 (1957).

Ō Murchadha, N., and York, J. W., Jr., "Initial Value Problem of General Relativity, I," *Phys. Rev. D* **10**, 428–436 (1974*a*).

———, "Gravitational Energy," *Phys. Rev. D* **10**, 2345–2357 (1974*b*).

Price, R. H., "Nonspherical Perturbations of Relativistic Gravitational Collapse I," *Phys. Rev. D* **5**, 2419–2438 (1972*a*).

———, "Nonspherical Perturbations of Relativistic Gravitational Collapse II," *Phys. Rev. D* **5**, 2439–2454 (1972*b*).

Schoen, R., and Yau, S. T., "The Energy and the Linear Momentum of Space-times in General Relativity," preprint, Institute for Adv. Study, Princeton, N.J. (1980).

Smarr, L., "Surface Geometry of Charged Rotating Black Holes," *Phys. Rev. D* **7**, 289–295 (1973).

———, "Gauge Conditions, Radiation Formulae, and the Two Black Hole Collision," in *Sources of Gravitational Radiation* (edited by L. Smarr, Cambridge Univ. Press, Cambridge, 1979), pp. 245–274.

Smarr, L., and York, J. W., Jr., "Radiation Gauge in General Relativity," *Phys. Rev. D* **17**, 1945–1956 (1978*a*).

———, "Kinematical Conditions in the Construction of Spacetime," *Phys. Rev. D* **17**, 2529–2551 (1978*b*).

Vonnegut, K., Jr., *The Sirens of Titan* (Delacorte Press, New York, 1959).

Wheeler, J. A., "Geometrodynamics and the Issue of the Final State," in *Relativity, Groups, and Topology* (edited by C. DeWitt and B. DeWitt, Gordon and Breach, New York, 1963), pp. 315–520.

Wilson, J., "Stellar Collapse and Supernovae," in *Sources of Gravitational Radiation* (edited by L. Smarr, Cambridge Univ. Press, Cambridge, 1979), pp. 335–343.

York, J. W., Jr., "Conformally Invariant Orthogonal Decomposition of Symmetric Tensors on Riemannian Manifolds and the Initial-value Problem of General Relativity," *J. Math. Phys.* **14**, 456–464 (1973).

———, "Covariant Decomposition of Symmetric Tensors in the Theory of Gravitation," *Ann. Inst. Henri Poincaré* **21**, 319–332 (1974).

———, "Kinematics and Dynamics of General Relativity," in *Sources of Gravita-*

tional Radiation (edited by L. Smarr, Cambridge Univ. Press, Cambridge, 1979*a*), pp. 83–126.

———, "Robin Boundary Conditions for Linear Second-order Elliptic Vector Equations," notes, Univ. of North Carolina, Chapel Hill, (1979*b*).

———, "Energy and Momentum of the Gravitational Field," in *Essays in General Relativity: A Festschrift for Abraham Taub* (edited by F. Tipler, Academic Press, New York, 1980).

Notes on Contributors

DIETER BRILL Professor Brill is Professor of Physics at the University of Maryland. He came to Maryland in 1970, having previously taught at Princeton and Yale. A native of Germany, he has received the Humboldt-Foundation Senior Scientist Award and has been awarded a medal from the Collège de France. His research has concerned a wide range of topics in general relativity including the positivity of mass, the topology of spacetime models, gravitational waves and perturbation theory.

SUBRAHMANYAN CHANDRASEKHAR Professor Chandrasekhar received both a Ph.D. and an Sc.D. from Cambridge University where he was a fellow at Trinity College before joining the faculty of the University of Chicago, where he has been since 1937. From 1952 to 1971 he was editor of the *Astrophysical Journal*.

ROY P. KERR Professor Kerr was Professor of Mathematics at the University of Texas at Austin from 1962 to 1971. Previously he had worked at Syracuse University and at the Aeronautical Research Laboratory at Wright-Patterson Air Force Base. Since 1971 he has been Professor of Applied Mathematics, University of Canterbury, Christchurch, New Zealand. He is noted for his work on exact solutions in general relativity. In particular, he derived the solution that bears his name, the Kerr metric for a rotating black hole, a solution of vast astrophysical importance.

CHARLES W. MISNER Professor Misner has visited the Texas Center for Relativity many times in the past at the invitation of Alfred Schild. He has worked in relativity theory since his graduate student work with John Wheeler at Princeton University in the 1950s. His research includes topological questions in spacetime geometries, mathematical cosmology, gravitational collapse, and gravitational radiation. He is well known as a coauthor (with K. S. Thorne and J. A. Wheeler) of the graduate text *Gravitation* and now works with an active research group at the University of Maryland.

TSVI PIRAN Professor Piran is a native of Israel. He received his Ph.D. from the Hebrew University of Jerusalem in 1977. While working on his contribution to this volume he was an Assistant Professor at the University of Texas at Austin, having previously been a Research Associate at Oxford University. He currently holds joint appointments at the Hebrew University and at the Institute for Advanced Study in Princeton.

IVOR ROBINSON Professor Robinson is Professor of Mathematics at the University of Texas at Dallas, having come to the predecessor of that institution, the Southwest Center of Advanced Studies, in 1963. Originally from England, he has lectured at Wales, North Carolina, Syracuse, and Cornell before coming to Texas. He has contributed important results to the study of gravitational radiation and the null paths along which photons and gravitons travel. The Bel-Robinson tensor, a geometric object important in gravitational radiation theory, is partly named after him.

DENNIS SCIAMA Professor Sciama is a Fellow of All Souls College, University of Oxford, where he is a leader of a research group in astrophysics. Before coming to Oxford, he had established a successful group in this subject at Cambridge University. He has published many technical papers on subjects ranging from black holes and quantum gravity to the formation of galaxies. He has also written three popular works on relativity and cosmology and is completing a book on black hole thermodynamics. Since 1978 Dr. Sciama has divided his time between Oxford and the University of Texas at Austin.

JAMES W. YORK, JR. Professor York is Professor of Physics at the University of North Carolina, Chapel Hill. Before going to Chapel Hill, he spent five years on the faculty of Princeton University and three years teaching at North Carolina State University, where he had begun the transition from engineering and applied physics to theoretical physics. He has published numerous technical papers concerning the initial value problem.

Index to Works Cited

Bold face characters indicate location of the full citation.

Index of Topics

Printed and bound by CPI Group (UK) Ltd, Croydon, CR0 4YY

27/10/2024

14580154-0001